Synthesis and Biomedical Applications of Multifunctional Poly(ethylene glycol) Derivatives

Zur Erlangung des akademischen Grades eines

DOKTORS DER NATURWISSENSCHAFTEN

(Dr. rer. nat.)

Fakultät für Chemie und Biowissenschaften
Karlsruher Instituts für Technologie (KIT) – Universitätsbereich
genehmigte

DISSERTATION

von

Master of Science Ekaterina Yur'evna Sokolovskaya

aus Dunaivtsi, Ukraine

Dekan:	Prof. Dr. Martin Bastmeyer
Referent:	Prof. Dr. Stefan Bräse
Korreferent:	Prof. Dr. Jörg Lahann
Tag der mündlichen Prüfung:	19.07.2013

Band 38
Beiträge zur organischen Synthese
Hrsg.: Stefan Bräse

Prof. Dr. Stefan Bräse
Institut für Organische Chemie
Karlsruher Institut für Technologie (KIT)
Fritz-Haber-Weg 6
D-76131 Karlsruhe

Bibliografische Information der Deutschen Bibliothek

Die Deutsche Nationalbibliothek verzeichnet diese Publikation in der
Deutschen Nationalbibliografie; detaillierte bibliografische Daten sind
im Internet über http://dnb.d-nb.de abrufbar.

ISBN 978-3-8325-3522-3
ISSN 1862-5681

Logos Verlag Berlin GmbH
Comeniushof, Gubener Str. 47,
10243 Berlin
Tel.: +49 030 42 85 10 90
Fax: +49 030 42 85 10 92
INTERNET: http://www.logos-verlag.de

Die vorliegende Arbeit wurde in der Zeit von April 2010 bis zum Juni 2013 am Institut für Funktionelle Grenzflächen des Karlsruher Instituts für Technologie (KIT) und an der University of Michigan, USA unter der Betreuung von Herrn Prof. Dr. Jörg Lahann und Herrn Prof. Dr. Stefan Bräse angefertigt.

Hiermit erkläre ich, dass ich die vorliegende Arbeit selbstständig und nur unter Verwendung der angegebenen Hilfsmittel sowie der zitierten Literatur angefertigt habe. Die Dissertation wurde bisher an keiner anderen Hochschule oder Universität eingereicht.

Table of Contents

1 Abstract

Rapid development of biomedicine creates a continuous demand for new polymers with tailored properties for the fabrication of various biomaterials. Poly(ethylene glycol) (PEG) is particularly suitable for this purpose due to its biocompatibility and exceptional protein resistance, however introduction of new functional groups as well as the increase of their number per polymer chain is required in order to tune PEG's properties and increase its loading capacity. The most efficient approach to do this is polymerization of functional epoxides and subsequent post-polymerization modification.

Current work consists of two major parts related to the preparation of multifunctional PEGs and their applications in the fabrication of biomaterials.

First, a general method for the anionic ring-opening polymerization of ethylene oxide and functional epoxides has been developed. Various polymerization conditions were investigated and compared. Under optimized conditions, a number of multifunctional PEGs and their copolymers with PEG have been prepared. The synthesized polymers were purified chromatographically on silica gel. In some instances, protection of a functional group was utilized during polymerization and removed thereafter.

Then, the obtained polymers were either directly or after post-polymerization modification utilized for the fabrication of biomaterials. Smart hydrogels have been prepared from a multifunctional photoresponsive PEG copolymer and multifunctional hydrazide PEG. Reactive anisotropic particles suitable for either reversible sugar conjugation or dual stimuli (oxidation and UV irradiation) triggered degradation have been fabricated via electrohydrodynamic co-jetting. Additionally, reactive surfaces were prepared by microcontact printing of a functional epoxide for further biomolecular conjugation.

These new biomaterials have great potential for applications in tissue engineering and targeted drug delivery with controlled release.

2   Introduction

Important advances have been made in recent years in the development of new biomaterials for applications such as drug delivery, tissue engineering and medical diagnostics.[1,2,3] Biomaterials have been defined as materials and devices that directly interact with human tissues and organs.[4] As the interaction occurs through their surfaces, it is vitally important to control the surface properties of biomaterials to ensure integration with host tissues. Surface attributes (such as wettability, charge and surface reactivity) depend on the chemical and physical characteristics of the molecular interface and have profound effects on biological responses.[5] In controlling material interactions in a biological environment, polymers are generally suitable as their properties can be greatly varied depending on the introduced functionalities, molecular weight and architecture.[4] Besides, it is often desirable that biomaterials show specific behavior in response to different stimuli, for example for controlled drug release or degradation of a scaffold in tissue engineering. This can be accomplished via introduction of stimuli-responsive groups into the polymer structure or creation of copolymers with different properties of the blocks.[6,7,8,9] Rapid developments in biomedical engineering create a continuous demand for polymers with newly designed properties for specific applications.

Various natural[10,11,12] and synthetic[13,14,15] polymers have been utilized to fabricate biomaterials. Among them poly(ethylene glycol) (PEG)* has received particular attention and has already demonstrated its great potential for biomedical applications due to its unique properties such as biocompatibility, high protein resistance, long circulation time in blood and lack of immunogenicity, antigenicity and toxicity.[16,17,18,19] It was approved by the Food and Drug Administration (FDA)[20] and has already been implemented into practice.[16,21]

In the following chapters, the most important areas of PEG applications are detailed and the key limitations connected with the use of commercially available PEG are presented. Possible approaches to PEG modification in order to overcome these limitations are further discussed.

---

* Polymers of ethylene oxide (EO) are also often refered to as poly(ethylene oxide) (PEO). Term PEG is commonly applied to polymers with molecular weight below 20,000 g/mol, while PEO – to those with molecular weight above 20,000 g/mol.

2.1 PEGs Application in Biomedicine

2.1.1 Surface Modification

Use of medical implants became a common practice in medical surgery for treatment of different diseases or injuries. Examples include coronary stents, heart valves, hip and knee prosthesis, cochlear implants, catheters, dental implants. Typically, non-natural materials such as metals or ceramics are used to produce implants in order to achieve desired mechanical strength.[22] Introduction of foreign objects into an organism causes a sequence of specific responses, including adsorption of plasma proteins, which can result in blood coagulation or activation of a cascade of protective mechanisms, such as attraction of destructive enzymes and production of hydrogen peroxide.[4] These are further accompanied by colonization of bacteria on the surface of an implant followed by biofilm formation which can finally require surgical removal of the implant.[23] To avoid undesired *in vivo* responses polymer coatings with specific properties have been employed to provide good integration of an implant with the body.

PEG-based films are commonly used as implant coatings due to the unique resistance of PEG to protein adsorption and cellular adhesion.[4,24,25] Star- and dendrimer-like PEGs showed better protein-repellant properties as compared to linear PEGs due to their compact structure and high segment density.[26] Additional properties such as antimicrobial characteristics or gradual release of a drug can be achieved via chemical modification of PEG end groups.[27] For this purpose, multifunctional PEGs are advantageous having a higher number of reactive groups suitable for modification.

Various techniques to introduce PEG films on surfaces have been developed including very simple methods such as spin coating,[28] spray coating[4] and dip coating[29] or more sophisticated ones such as plasma deposition.[23] However to obtain highly stable coatings, covalent binding of PEG to the substrate is preferable. A disadvantage of common covalent grafting is a relatively low density of the obtained polymer film, which affects the protein resistance properties, therefore this method mostly utilizes branched PEG structures.[29] The high surface density of PEG chains can be achieved by formation of PEG self-assembled monolayers (SAM) on various substrates. For example, SAMs on silicon surfaces can be formed by using silanated PEG,[30,31] while SAMs on gold covered substrates can be prepared with PEG-thiols.[32] The method is easy in application and allows for thermodynamic control resulting in highly organized structures with equally distributed end functional groups of PEG. Since SAM formation is typically performed in a water solution and at room temperature, various chemical functionalities for further modification can be introduced this way, which might not be stable when using other surface modification methods performed in harsher conditions.[33] The limitation for long term use of SAM coatings is their low robustness since the coating is a single molecular layer, and appearance of the defects would result in decreased protein resistance. Surface-induced polymerization of ethylene glycol oligomers does not have this limitation and allows the achievement of high PEG surface density combined with the possibility of tuning the thickness of the layer and architecture of the introduced polymer. This was realized for example by surface-initiated atom transfer radical polymerization (ATRP) of oligo(ethylene glycol) methyl methacrylate (OEGMA) to form PEG brushes on gold surfaces.[34] PEG coatings were also introduced on other substrates besides silicon or gold. For example, metal surfaces were modified by using PEG conjugates with 3,4-dihydroxyphenylalanine (DOPA)[35] or cyanobacterial iron chelator anachelin.[36]

Functional PEGs can be further crosslinked after immobilization on a surface with the use of suitable crosslinkers or by photocuring resulting in surface hydrogel formation. Isocyanate terminated PEGs are often utilized for chemical crosslinking via reaction with amines or alcohols.[29,37] Crosslinking of a PEG coating increases its stability and density and results in the formation of a more homogeneous layer and, as a consequence, very high resistance to protein adsorption and cells adhesion.

2.1.2 Drug Delivery

Drug delivery is another important and deeply studied area of PEG applications in biomedicine. Conjugation of a therapeutic molecule to PEG is a well-known approach used in drug delivery termed "PEGylation".[16] PEGylation can be applied to either small molecular weight drugs or proteins. This approach improves the properties of the conjugate such as prolonged circulation half-life in body, protection against degrading enzymes, decreased immunogenicity, and in case of some small molecular drugs – increased water solubility.[16] Several PEGylated proteins have already been approved for medical use by the FDA, such as PEG-L-asparaginase (Oncospar; Enzon) for treatment of acute lymphoblastic leukemia,[38] PEG-G-CSF (Neulasta, Amgen) – for treatment of neutropenia during chemotherapy,[39] PEG-

Interferon α-2b (PegIntron, Schering) – for treatment of chronic hepatitis C and hepatitis B,[40] which in 2011 was also approved for treatment of melanoma under the brand name Sylatron (Merck).[41] Many others are currently involved in clinical trials at different stages.[16,42] While most success was achieved with PEGylated proteins, PEG conjugates with small molecules have a crucial limitation for the use in drug delivery resulting from the low loading capacity of the commonly used PEG-diol and as a consequence disadvantageous drug to polymer weight ratio.[16] Star- and dendrimer-like PEGs are more promising for the delivery of low molecular weight drugs as they have an increased number of functional groups suitable for the drug-molecule conjugation.[26]

PEGs have also been successfully utilized to create micelles and nanoparticles as drug carriers which were shown to be more efficient than conventional PEG-conjugates due to the specific properties of such carriers arising from their nanoscale size and surface characteristic.[21] PEG-based micelles can be prepared from block copolymers of PEG with hydrophobic polymers, due to their ability to self-assemble in water solutions above the critical micelle concentration (CMC).[43] The hydrophobic block forms the core of a micelle, while hydrophilic PEG block composes the shell. Due to the formation of this structure hydrophobic drug molecules can be encapsulated into the core part of a micelle and successfully delivered to the targeted tissues. The examples of PEG copolymers able to form micelles include those with poly(propylene oxide) (PPO)[44] phospholipids,[45] poly(amino acids),[46] poly(esters),[47] etc.

Furthermore, anisotropic particles may enhance the efficiency of drug delivery by additional conjugation of targeting ligands,[48] or combining delivery with imaging,[49] or allowing for multiple delivery of drugs with different release kinetics.[50] The use of polymeric particles for drug delivery will be discussed in more details in Chapters 4.2.3 and 4.2.4.

2.1.3 Tissue Engineering

Materials suitable for use in tissue engineering applications should provide an environment similar to that of a human organism to ensure cell proliferation and integration with natural tissues. Hydrogels are mostly suitable for this purpose due to their highly hydrophilic nature and mechanical properties. Among available synthetic polymer building blocks for the hydrogel formation ethylene glycol-based polymers are particularly attractive thanks to their hydrophilicity and biocompatibility.[24] Swelling ability and elasticity of PEG gels can be varied by using polymers with different architectures and branch lengths as well as by the use of PEG copolymers.

Triblock copolymer of EO and propylene oxide (PO) PEO-PPO-PEO commercialized under the trade name "Plurionic" is one of the PEG materials commonly used to produce hydrogels for tissue engineering.[51] It undergoes a thermoreversible gelation at close to body temperatures, which makes *in situ* formation of hydrogels upon injection possible. For example, Pluronic was used to encapsulate chondrocytes to produce artificial cartilage.[52] Physical gels formed by Pluronic readily degrade upon dilution with body liquids.[53] In order to provide a better stability of PEG-based hydrogels crosslinking of PEG chains is usually performed. Most commonly this is realized by photocrosslinking of PEG acrylate derivatives, since the reaction proceeds in mild conditions and can be directly performed in the presence of cells allowing for an efficient encapsulation.[54] Besides, various click reactions have been utilized to perform gentle crosslinking which would not interfere with cells in the scaffold.[55,56]

Though PEG alone is not biodegradable, the degradability of its hydrogels can be ensured via introduction of degradable crosslinking units or by utilization of PEG copolymers with degradable poly(lactic acid) (PLA) or poly(glycolic acid) (PGA).[57]

To provide a better environment for cell proliferation and formation of tissues, the properties of a hydrogel should be as close as possible to the natural environment of extracellular matrix (ECM). Since PEG is a bio-inert polymer, hydrogel matrix modification with biomolecules which present in ECM such as proteins and glycans have been performed to ensure biofunctioning and desired responses of the cells cultured in such matrix. These dynamic hydrogels were used to encapsulate cells to provide bioactive scaffolds for tissue engineering.[57]

Insufficient cell adhesion is one of the major problems connected with engineered hydrogel scaffolds. To overcome this limitation, RGD (Arg-Gly-Asp) – a peptide derived from an ECM protein – is commonly used for immobilization on PEG hydrogels. Lee et al. investigated RGD-modified crosslinked Pluronic-based hydrogel as a culture matrix for chondrocytes.[58] Presence of RGD provides the desired interaction of the hydrogel with chondrocytes, while temperature dependence of the gel properties allows for *in situ* gel formation upon injection.

This short overview of the most important areas of PEGs applications reveals the necessity of PEG functionalization. Commercially available PEG and its methoxy analogue (mPEG) have just two and one hydroxyl groups respectively that are suitable for further modification. However presence of functional groups is essential, on one hand, for the variation of physical properties of PEG via chemical modification, and on the other, for increasing its loading capacity in drug delivery, crosslinking to form hydrogels and biomolecular conjugating to promote cells proliferation on surfaces. The attempts to overcome this shortage have resulted in the synthesis of different bi- and multifunctional PEGs with various architectures.

2.2   Towards Multifunctional PEGs

2.2.1   Bifunctional PEGs

Several important applications have urged the development of new efficient synthetic methods to obtain α,ω-heterobifunctional PEGs, including the discussed above surface modification (Chapter 2.1.1), drug delivery (Chapter 2.1.2) and tissue engineering (Chapter 2.1.3). Furthermore, if both functional groups of a α,ω-heterotelechelic PEG are reactive and capable to undergo orthogonal chemistries, then such bifunctional PEGs can serve as linkers, for example, to bind bioactive molecules[59] or to perform surface-initiated polymerization.[60]

Synthesis of homobifunctional or α,ω-heterotelechelic PEGs can be done using one of the following approaches:[61]

- direct approach: use of suitable initiators and terminating agents bearing functional/protected functional group for EO polymerization (Scheme 1A),

- indirect approach: modification of terminal hydroxyl groups of commercially available PEG-diol (Scheme 1B).

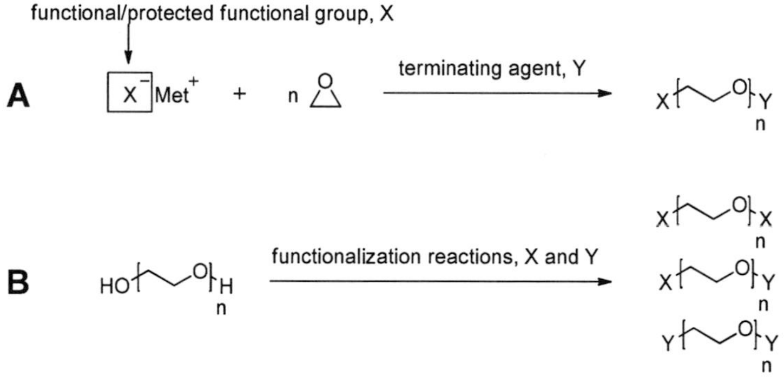

Scheme 1   Approaches to the synthesis of heterobifunctional PEGs: direct approach via EO polymerization with the use of functional initiators and terminating agents (A), indirect approach via modification of terminal hydroxyl groups of PEG-diol (B).[61]

Synthesis of Heterobifunctional PEGs via EO Polymerization with the Use of Functional Initiators and Terminating Agents (Direct Approach)

The direct approach involves polymerization of EO, which is commonly performed as a living anionic ring-opening polymerization and results in a quite narrow molecular weight distribution of the obtained polymer.[61] Hydroxyl anion or an alkoxide are typically used to initiate the polymerization. For example, mPEG is prepared by EO polymerization initiated by potassium methoxide.[62] Instead, an alkoxide bearing functional/protected functional moiety compatible with the polymerization conditions can be used to introduce other functionalities. So far the groups directly introduced into EO polymerization without the use of protection are alkene, carboxyl and cyano groups.

The alkene moiety is most often introduced by initiation with allyl alkoxide,[63] although variations are possible.[61,64,65] For example, potassium trivinylsilyl propialate used as initiator not only introduced an alkene moiety at one of the PEG's termini, but also allowed an increase in the number of functional groups per polymer molecule (Scheme 2).[61] Introduction of terminal double bonds into PEG chain is valuable due to their specific reactivity toward thiols,[66] which opens further possibilities for modification. Using thiol-ene reaction conversion of terminal vinylsilane groups into carboxylic acid groups was successfully performed (Scheme 2).

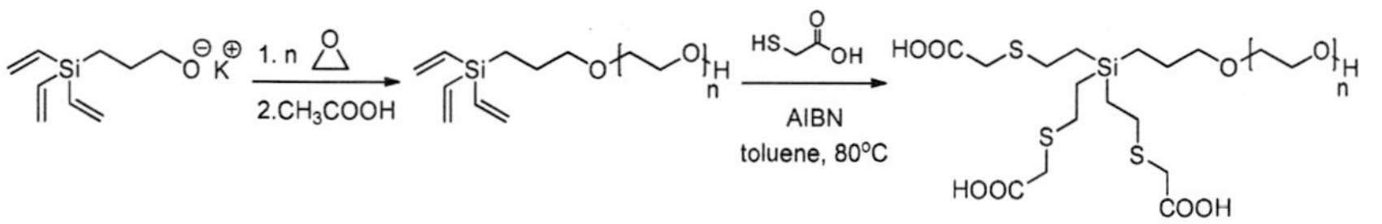

Scheme 2 Polymerization of EO using a trivinylsilyl propialate initiator and subsequent modification of the obtained (allyl)$_3$–PEG–OH into the corresponding (HOOC)$_3$–PEG–OH.[61]

Carboxy-functionalized heterobifunctional PEG can also be synthesized by direct polymerization of EO initiated by potassium thiolate bearing a carboxylic acid group. Zeng and Allen have evaluated different oxy-anionic and thio-anionic functional initiators with protected and unprotected carboxylic acid groups.[67] The authors showed that dipotassium 3-mercaptopropionate can efficiently initiate EO polymerization yielding the desired bifunctional PEG with carboxyl and hydroxyl groups at the termini. This is possible due to the significant difference of the reactivities of thiolate and carboxylate ions, which prevents the interference of the latter with polymerization process and allows for complete recovery of the acid moieties after termination of polymerization (Scheme 3).

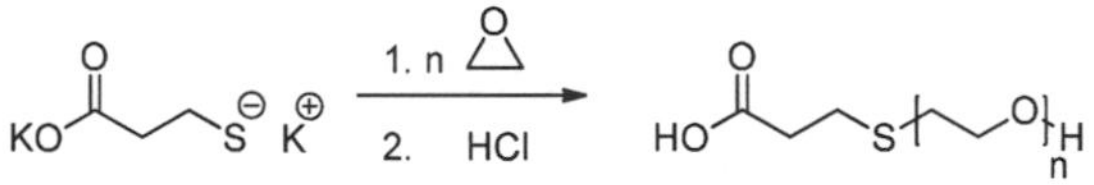

Scheme 3   Polymerization of EO using a thiolate initiator with carboxylic acid group.[67]

Polymerization of EO initiated by cyanomethyl potassium introduces cyano group at one of the PEG's termini.[68] This group can either be further hydrolyzed into a carboxylic acid group[68] or reduced into amino group.[69]

Although many functional groups undergo side reactions in highly basic conditions of anionic polymerization, the use of suitable protections can be a possible solution to overcome this limitation. For example, an amino group can be protected via imine formation.[70] In another example, trialkylsilyl protection for amino groups was utilized.[71] The authors initiated EO polymerization with commercial potassium bis(trimethylsilyl)amide (KHMDS). Quantitative cleavage of trimethylsilyl groups was achieved by acid treatment. However, the authors detected some chain transfer reactions due to the low resistance of trimethylsilyl protecting group against nucleophilic potassium alkoxide. To avoid any possible side reactions, a more stable trialkylsilyl protection was employed to obtain PEG with terminal amino and hydroxyl groups (Scheme 4).[72]

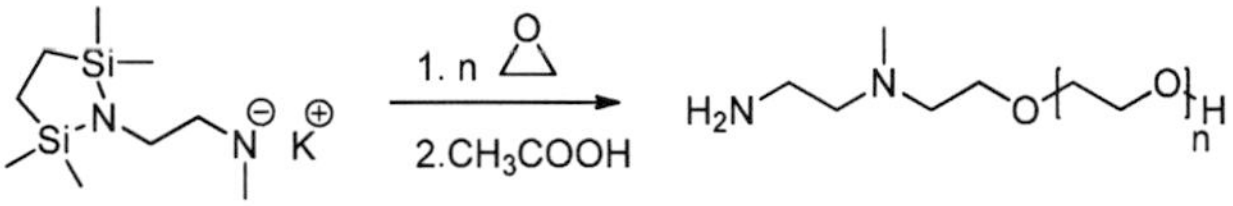

Scheme 4   Polymerization of EO using an initiator with trialkylsilyl protected amino group.[72]

Acetal bonding is stable under basic conditions, therefore it was utilized as a protection for aldehyde group during EO polymerization (Scheme 5).[73,74,75] The aldehyde moiety is easily recovered by acid treatment.

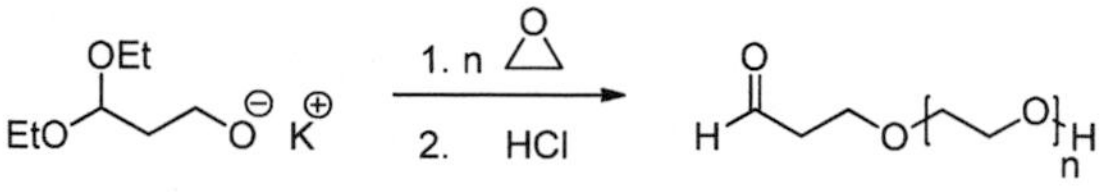

Scheme 5   Polymerization of EO using an initiator with acetal protected aldehyde group.[73]

Acetal protection can also be applied to a hydroxyl group. Commonly used in organic synthesis to protect alcohols, tetrahydropyranyl (THP) protection was successfully employed in EO polymerization allowing for quantitative differentiation of PEG's terminal hydroxyls (Scheme 6).[76] After post-polymerization modification of the free hydroxyl, THP protection was cleaved in acidic conditions releasing the second hydroxyl, which can further be converted into other functional groups.

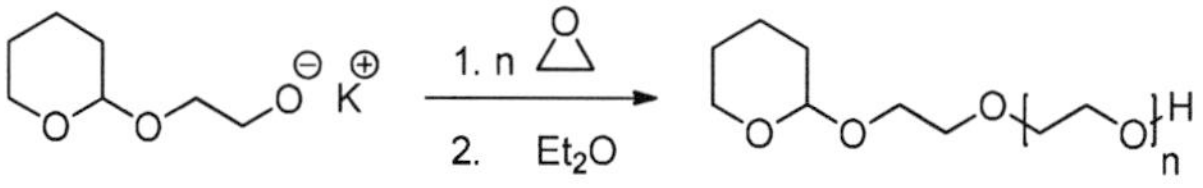

Scheme 6   Polymerization of EO using an initiator with acetal protected hydroxy group.[76]

Another type of acetal protection for hydroxyl, namely 1-ethoxyethoxy group, was utilized by Li and Chau.[64] In addition, the use of benzyl protection for hydroxy and amino groups in EO polymerization was reported.[61,77] Cleavage of benzyl protection by hydrogenolysis yields free hydroxyl and amino groups correspondingly.

Finally, disulfide moiety – a precursor of a thiol group, was introduced at one of the PEG's termini with initiator during EO polymerization (Scheme 7).[78] This disulfide group can be easily converted into a corresponding thiol by reduction.

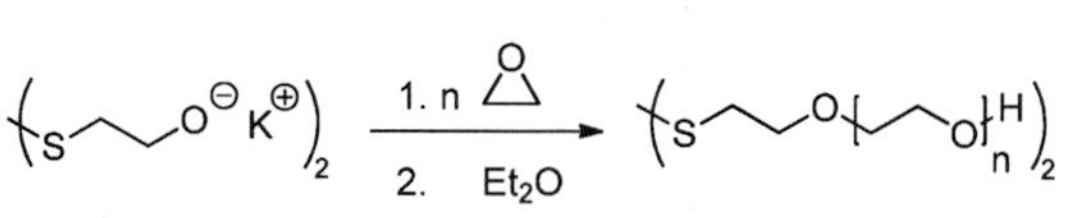

Scheme 7   Polymerization of EO using an initiator with disulfide group.[78]

Due to the living nature of ring-opening polymerization, quantitative termination with functional agents is possible. Thus besides hydroxyl groups which are derived by addition of acids to the living PEG chains, other functionalities can be introduced. Since there are less limitations to the groups introduced with terminating agents as compared to those introduced with initiators under harsh anionic polymerization conditions, a wider range of functionalities can be integrated by the former method. Termination of polymerization is commonly performed by addition of nucleophile reactive acid derivatives (acid anhydrides, acid chlorides, succinimidyl esters, etc.) or halogenides.[61]

A special metal free variant of EO polymerization was realized with the use of N-heterocyclic carbene, 1,3-bis-(diisopropyl)imidazol-2-ylidene, as an initiator (Scheme 8).[79,80] Polymerization proceeds by a zwitterionic ring-opening mechanism and allows for an easy introduction of different functional groups at one of the PEG's termini and hydroxyl group – at the other by termination with a suitable functional agent. The advantage of this method is metal free environment of the polymerization, preventing the presence of metal traces in the produced bifunctional PEG, which is important for further biomedical application.

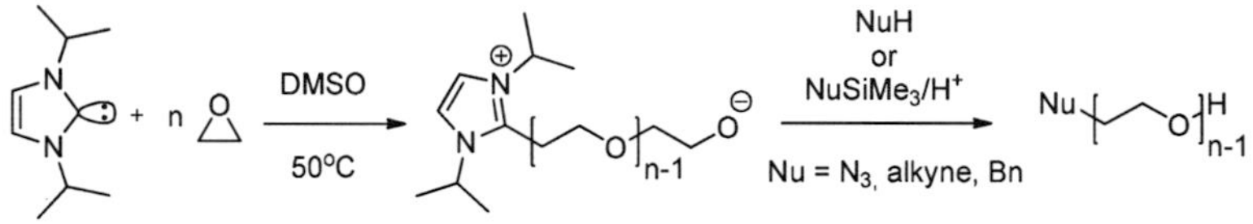

Scheme 8   Polymerization of EO using a metal free N-heterocyclic carbene initiator.[79]

The use of functional initiators and terminating agents during EO polymerization allows for a quantitative introduction of new functional groups at the PEG termini, due to which complicated and time-consuming procedures of product purification can be avoided. However, special care should be taken to remove any water traces from the polymerization media, which can cause termination reactions and initiate EO polymerization resulting in the formation of PEG-diol.

Synthesis of Heterobifunctional PEGs via Modification of Terminal Hydroxyl Groups of PEG-Diol (Indirect Approach)

Alternatively, an indirect approach to synthesize heterobifunctional PEGs that relies on alternation of terminal PEG's hydroxyl groups can be utilized (Scheme 1B).[61] Quantitative differentiation of two identical hydroxyls is impossible and results in the formation of a mixture of products consisting of mono-, di- and non-modified species, which require further separation. Since the properties of PEGs differed by just one or two terminal groups are quite similar, the separation is a rather complicated procedure and commonly results in low yields of heterobifunctional PEG. Furthermore, with the increase of molecular weight the difference in properties of mono-, di- and non-modified PEGs decreases making separation even more complicated. As a result, this approach is mostly applied for the synthesis of low molecular weight bifunctional PEGs.[61] Separation of PEGs bearing ionizable functional groups such as amine or carboxylic acid is possible by ion exchange chromatography.[81] Column chromatographic purification on silica gel is another available technique to separate heterobifunctional PEGs from the reaction mixture. This method was successfully applied for the purification of commercial mPEG from the traces of PEG-diol.[82] The content of PEG diol in the purified mPEG was less than 1%. However chromatographic purification on silica gel can efficiently be applied just in case of low molecular weight PEGs.

Recently, Mahou and Wandrey have reported on the selective monotosylation of commercial PEG-diol in the presence of silver oxide and a catalytic amount of potassium iodide (Scheme 9).[83]

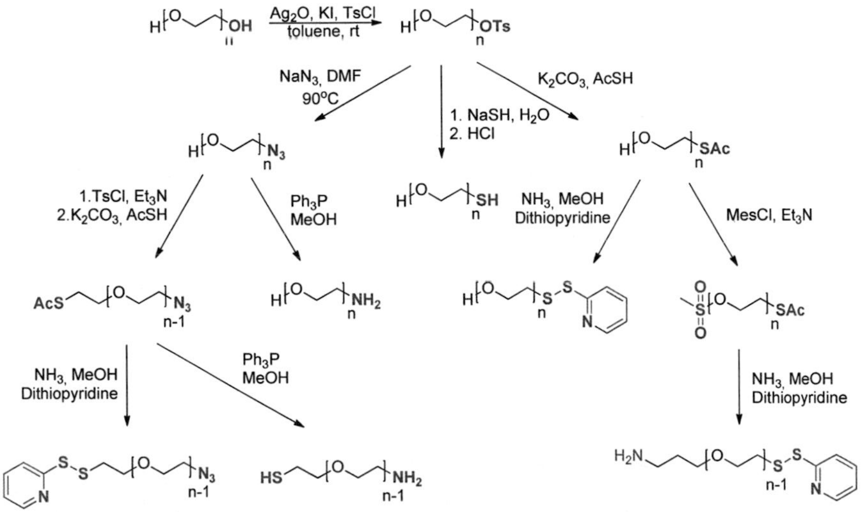

Scheme 9   Synthesis of a library of heterobifunctional PEGs starting from its monotosylated derivative.[83]

The authors estimated the degree of functionalization as almost quantitative (~99%) without detected amounts of PEG-diol. They suggested that the selectivity of the reaction can be explained by intramolecular hydrogen bond formation, resulting in the decreased acidity of the involved hydrogen and selective deprotonation of the other hydroxyl group by silver oxide. Monotosylated PEG derivative was used as a precursor for the synthesis of a library of different heterobifunctional PEGs.

Though modification of PEG terminal hydroxyl groups in the majority of cases is complicated by the necessity to separate the product from a mixture of bifunctional PEGs, resulting in considerably lower yields of the target polymer as compared to those obtained by EO polymerization with the use of functional agents, it has several advantages. Specifically, it does not require the use of toxic EO and is tolerant to a broader range of functional groups.

2.2.2 Branched PEGs

Introduction of new functional groups at PEG's termini widens the range of its applications allowing for conjugation of therapeutic molecules via various linkages as well as for modification of different functional surfaces. However this does not improve the low loading capacity of PEG, which is crucial for the delivery of low molecular weight drugs. Synthesis of PEGs with different branched architectures is a feasible solution to overcome this drawback, since the number of terminal groups suitable for conjugation increases considerably. Furthermore, branched structures allow for better encapsulation of a drug preventing it from the destruction in the organism. These factors make branched PEGs more attractive carriers as compared to linear bifunctional PEGs. Furthermore, branched PEGs have much higher capacity for hydrogel formation (Chapter 2.1.3) and form more compact structure on surfaces, which consequently exhibit better protein resistance (Chapter 2.1.1).

A comprehensive overview of the strategies towards star-like PEGs synthesis as well as the range of thus far available structures was made by Lapienis.[26] These approaches are schematically presented in Scheme 10 and can be classified as the following:[26]

- "core first" approach implicates polymerization of EO initiated by multifunctional molecules. Such initiators can either have a definite number of initiating centers (for example, polyols) or can themselves be a branched or crosslinked structure with

randomly distributed groups suitable for initiation of EO polymerization
(plurifunctional core) (Scheme 10A),

-   "arm first" approach uses a preformed linear PEG, which reacts via its activated (can
    be living) terminal group with multifunctional molecule, which as in the case of the
    first approach can have a defined or plurifunctional structure. If the core molecule has
    only one type of functional groups then homostar polymer is obtained. In the presence
    of two or more different types of reactive groups heteroarm star can be synthesized
    (Scheme 10B),

-   "in-out" approach combines the two previous methods and first involves conjugation
    of a preformed linear PEG to the core molecule via reaction with a part of functional
    groups of the latter followed by the activation of the rest of its functional groups for
    initiation of EO polymerization. (Scheme 10B).

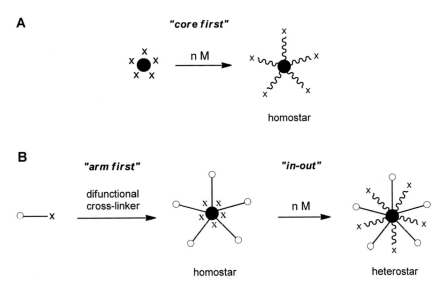

Scheme 10   "Core first" (A) and "arm first" (B) approaches to the synthesis of star-like
                PEGs.[26]

Dendrimer-like PEGs are well-defined structures synthesized by a stepwise modification
method (Scheme 11).[84,85] The synthesis starts from a core unit with multiple functionalities
which adds PEG branches. This addition may come from the reaction with one of the linear
PEG's terminal groups (analogous to "arm first" approach for the synthesis of star-like PEGs)

or can also be realized via EO polymerization initiated by the multifunctional core molecule (analogous to "core first" approach). This step enables the production of the first generation dendrimer-like macromolecule. Next, two or more reactive groups are introduced at the arm ends which serve as branching points in the architecture. The following branching molecules were used in dendrimer-like PEG synthesis: 2-amino-1,3-propandiol,[86] amino adipic acid,[87] 3,5-dioxybenzoate,[88] allyl chloride/OsO$_4$,[89] etc.[26] The addition of a second PEG layer from these branching points enables the formation of the second generation of the dendrimer-like polymer. Higher generation dendrimer-like PEGs are produced by repetition of this process. Branching can also be performed via formation of degradable linkages such as urethane bonds to create biodegradable polymers.[90] In case of EO polymerization branching units can be introduced by termination with agents bearing two or more functional group which can be further activated to perform next step polymerization.[26]

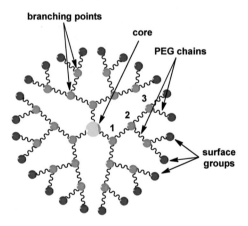

Scheme 11   Schematic representation of a dendrimer-like PEG structure.

Dendrimer-like PEGs prepared by an "arm first" technique have a precise architecture and low polydispersities since there is no broadening of molecular weight during the modification steps and polydispersity of the final polymer is defined by the polydispersity of the utilized linear PEG.

Unfortunately, synthesis of dendrimer-like PEGs is quite a time- and labor-consuming procedure and is less efficient for high generation dendrimer-like polymers because of the high density of functional groups, resulting in their incomplete modification and appearance of defects in the dendrimeric structure.

2.2.3 Multifunctional PEGs

Though synthesis of branched PEGs allows for a considerable increase in the number of functional groups per PEG molecule as compared to linear PEGs, the functionalities introduced are still located just at the polymer termini. To introduce multiple functionalities into the polymer backbone and also to access multifunctional linear PEGs, polymerization of substituted epoxides has to be performed (Scheme 12A). The density of functional groups per polymer chain increases dramatically as compared to branched PEGs and can be controlled by copolymerization with EO using a defined comonomers feed ratio (Scheme 12B). Finally, copolymerization of different functional epoxides can be performed in order to obtain heteromultifunctional polymers (Scheme 12C). Thus a designed PEG-based polymer with domains having specific and independent functions (given physical properties, biomolecular conjugation, crosslinking, surface reactions, stimuli-response, etc.) can be accessed (Scheme 12D). Additional functionalization at the PEG termini is possible via the use of functional initiators and terminating agents (Chapter 2.2.1).

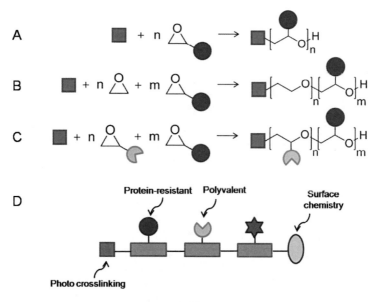

Scheme 12   Schematic representation of the synthesis of multifunctional PEGs by (co)polymerization of functional epoxides.

Not all functional groups are compatible with the polymerization conditions, especially with the strong basic environment of anionic polymerization (Chapter 2.2.1). Sometimes coordination polymerization can be a solution since it proceeds in milder conditions.[91] In general, the following strategies should be considered when planning synthesis of multifunctional PEGs (Scheme 13):[20]

-   direct polymerization of an epoxide with the desired functional group (Scheme 13A),
-   polymerization of an epoxide with protected functional group followed by post-polymerization cleavage of protection (Scheme 13B),
-   post-polymerization modification of the groups introduced via the first two strategies (Scheme 13C).

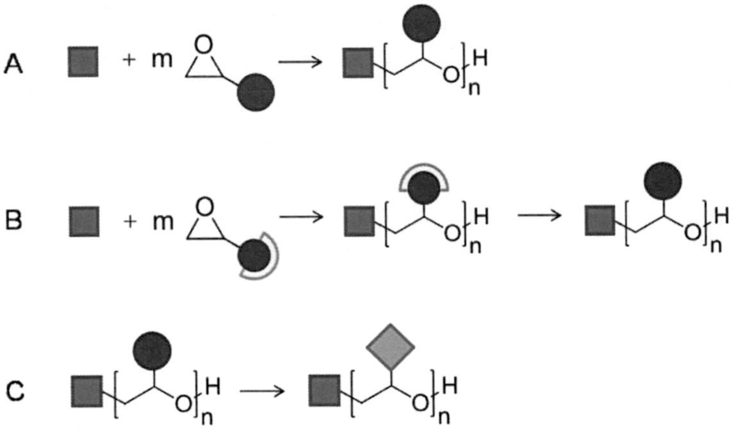

Scheme 13   Schematic representation of the synthesis of multifunctional PEGs by polymerization of functional epoxides.

The range of epoxides reported thus far for the synthesis of multifunctional PEGs and further discussed here is shown in Scheme 14. Besides, the polymerization of various alkyl and phenyl substituted epoxides has been intensively studied.[92,93,94,95] Although those epoxides are important for variation of physical properties of PEG, such as hydrophilicity, they are not discussed here since the introduced groups are inert and could not be used for further modification or conjugation of therapeutic molecules.

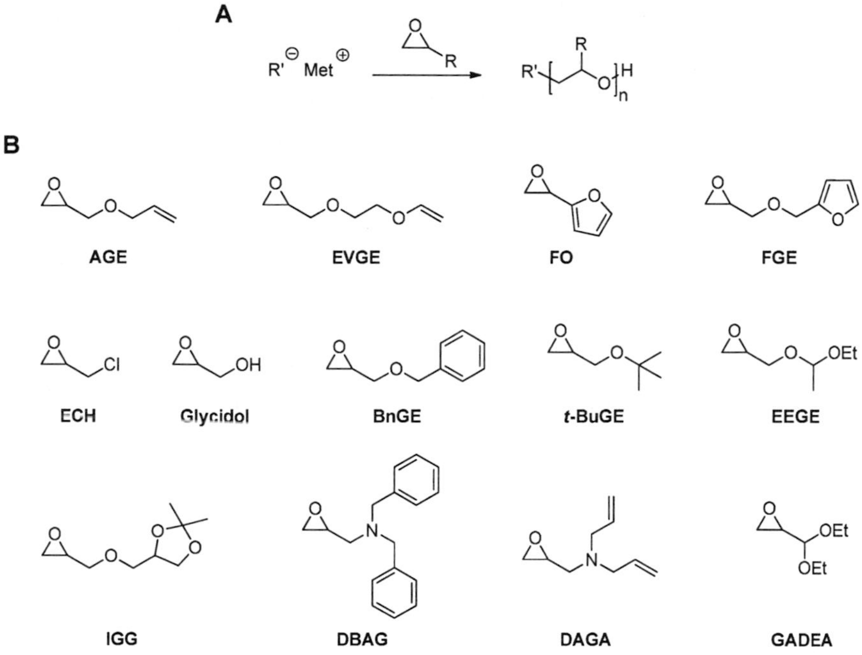

Scheme 14   Polymerization of functional epoxides (A) and the range of functional epoxides reported for the synthesis of multifunctional PEGs (B).

Direct Polymerization

A double bond can be directly introduced via anionic polymerization of allyl glycidyl ether (AGE) or ethoxy vinyl glycidyl ether (EVGE). Polymerization of AGE and its copolymerization with EO has been intensively studied.[96,97,98,99] The interest in PAGE stems from the possibility for further modification of the polymer via thiol-ene click reaction.[98,100]. Additionally, PAGE can be considered as a precursor of polyglycidol (PG), which can be accessed through isomerization of allyl to propenyl groups followed by cleavage in acidic media.[99] PEVGE is also suitable for thiol-ene modification, but in contrast to PAGE does not require excess of thiol, since the vinyl bond is inert to radical crosslinking reaction.[101] Furthermore, vinyl ether groups readily react with alcohols forming labile acetal bonding, which has potential applications in drug delivery with controlled release.

2-Furyloxyrane (FO) is also known to undergo anionic polymerization yielding corresponding multifunctional PEGs with furyl moieties.[102] It was shown that addition of crown ethers during polymerization is necessary in order to obtain high molecular weight PFO. The furyl group can also be introduced into the PEG chain via coordination polymerization of furfuryl glycidyl ether (FGE), which is often performed as copolymerization with carbon dioxide to access polycarbonates.[103,104] This furyl group can be utilized for reversible Diels-Alder click reaction with maleimide.[105] However, different reports indicated low stability of polymers with the furan ring, because of the tendency of the latter to crosslink.[104]

Epichlorohydrin (ECH) is an important commercially available (in both pure enantiomeric forms) epoxide for the synthesis of multifunctional PEGs suitable for subsequent modification by substitution of chlorine[106] and for the preparation of elastomeric polymers with good resistance properties.[107] Although conventional anionic polymerization of ECH is not possible, its polymerization via cationic and coordination mechanism has been deeply studied.[96,108] Recently, the use of tetraoctylammonium bromide-triisobutylaluminum initiating systems has been applied for copolymerization of ECH with other functional epoxides.[106,109]

Besides the examples described above, there are just few other cases of direct anionic polymerization of epoxides with some very specific groups, which have not yet found general applications.[110,111,112]

Use of Protecting Groups

For some functional groups incompatible with the polymerization conditions, suitable protecting groups can be found. The protection should be stable during polymerization and quantitatively cleaved thereafter.

PG is an important polymer which can be a promising substituent for non-functional PEG due to its high hydrophilicity and biocompatibility and has already been approved by FDA for the use as food and pharmaceutical additives.[113] Due to the presence of multiple hydroxyl groups in its backbone, PG has an advantage over PEG-diol for further modification or drug loading. Direct polymerization of glycidol leads to the formation of hyperbranched structures.[114] In order to obtain linear polymer hydroxyl group of glycidol has to be protected prior to polymerization. It has been mentioned above that AGE can be considered as a protected glycidol. Besides, benzyl,[115,116] *tert*-butyl[99,115] and ethoxy ethyl[77] groups have also been utilized to protect hydroxyl group of glycidol. Due to the different conditions required for the cleavage of these protecting groups, selective stepwise deprotection and modification of hydroxyl groups in PG is feasible. The possibility to perform orthogonal cleavage of hydroxyl protecting groups in copolymers of AGE, *tert*-butyl glycidyl ether (*t*-BuGE) and ethoxyethyl glycidyl ether (EEGE) was studied by Möller and co-workers.[99] EEGE is most commonly utilized for the synthesis of PG due to the ease of acetal protection cleavage and quantitative conversions of PEEGE into PG.

Recently Frey and co-workers have reported on the polymerization of an EEGE analogue bearing two protected hydroxyl groups: 1,2-isopropylidene glyceryl glycidyl ether (IGG).[117] Correspondingly, PIGG has double the number of protected hydroxyl groups within PEG's backbone as compared to PEEGE. Furthermore, after deprotection glyceryl groups can be used to conjugate therapeutic molecules bearing aldehyde groups via acetal formation for further controlled drug release.

Formation of an acetal can also be used for protection of aldehyde groups. However, thus far just one work on coordination polymerization of glycidaldehyde diethyl acetal (GADEA) has been reported.[118]

The amino group is another important functionality for bioconjugation via formation of peptide bonds. In order to perform polymerization of an epoxide bearing an amino group, the latter has to be protected. The benzyl group is suitable for this purpose and can be cleaved by hydrogenolysis. A multifunctional PEG with protected amino groups in the polymer backbone was synthesized by copolymerization of EO with N,N-dibenzyl amino glycidol (DBAG).[119] However, the heterogeneous nature of the P(EG-*co*-DBAG) deprotection process prolonged the reaction time and resulted in low yields of the corresponding polyaminoglycidol P(EO-co-AG). Besides, synthesis of PEG-*b*-PAG block copolymers was not possible. These problems were overcome with use of N,N-diallylaminoglycidol (DAAG).[120] Synthesized P(EO-co-DAAG) was quantitatively converted into the corresponding amino multifunctional P(EO-co-AG) by cleavage of allyl protecting groups via isomerization into vinyl groups followed by subsequent acidic hydrolysis.

Post-Polymerization Modification

In the case of more complex functional groups, a suitable protection may not be found. Therefore, the only way to access these groups is post-polymerization modification of multifunctional polymers synthesized via the first two strategies (Scheme 13C). Reactions applied to the modification of polymers should proceed in quantitative and specific manner to avoid presence of different species within one macromolecule.[121] Therefore optimally click reactions should be employed.[122] As it has already been mentioned above PAGE can be modified via thiol-ene click reaction. This polymer and its copolymers were often used as precursors for the syntheses of various multifunctional PEGs with amino, carboxyl and hydrazide groups, sugars, etc.[98,100,123,124,125,126]

Other strategies for post-polymerization modification of multifunctional PEGs mostly rely on the reaction of PECH[106,127,128,129] and PG.[99,113,130] Li and Chau have synthesized a diverse library of homomultifunctional PEGs and their copolymers based on PG modification.[131]

PAG is a recently prepared polymer and the possibilities for its modification have not yet been deeply investigated.[119,120] Still it can also contribute in future to the development of the synthetic platform for the synthesis of various multifunctional PEGs.

Finally, PFO and PGADEA have great potential for the synthesis of new multifunctional PEGs by post-polymerization modification approach since the modification can be performed via Diels-Alder and aldehyde-hydrazide click reactions correspondingly. However, their synthesis by polymerization of FO and GADEA as well as their reactivity have not yet been well investigated and require additional study.

Based on the analyzed literature data, polymerization of functional epoxides is the most efficient way to access multifunctional PEGs since not only can a variety of new functionalities be introduced into the PEG molecule, but also their high density per polymer chain can be achieved and controlled since functional groups are introduced into the polymer backbone and not only at its termini as in the case of bifunctional linear PEGs and branched PEGs. However, the range of functional groups compatible with anionic ring opening polymerization and utilized thus far is limited and should be further investigated. Alternatively, the possibility of using suitable protecting groups as well as methods for quantitative post-polymerization modification should be further developed.

3   Scope of the Project

Advances in biomedicine create a continuous demand for new polymers with specifically designed properties for fabrication of biomaterials. Various biocompatible synthetic polymers have been prepared to satisfy this need among which PEG is keeping the leading positions. Still the lack of functional groups in the main PEG chain limits its flexibility and the scope of its applications. The most efficient way to overcome this shortage is polymerization of functional epoxides. However, thus far polymerization of only AGE and glycidol has received considerable attention and the corresponding polymers have been further employed for the preparation of biomaterials. Further development of this area and broadening of the library of available multifunctional PEGs with designed properties is required.

The aim of the current work is the synthesis of different multifunctional PEGs and their further applications for the fabrication of smart biomaterials.

Investigation of the synthetic conditions for polymerization of EO and its analogues is presented. The literature data on the polymerization of functional epoxides are very diverse and employ different reaction conditions and initiating systems. Moreover, chain transfer and other side reactions can take place during polymerization and lead to low quality polymers. Therefore a general method for the synthesis of multifunctional PEGs by living anionic ring-opening polymerization has been developed. For this purpose a range of functional epoxides has been prepared and (co)polymerized, including those not reported previously. Furthermore, additional possibility for functionalization via initiation with functional alkoxides has been studied.

Multifunctional PEGs as well as functional epoxides alone have a great potential for various biomedical applications (Scheme 15). In this study some of these applications have been investigated, including surface modification with reactive epoxides for further bioconjugation and fabrication of "smart" microparticles and hydrogels based on stimuli-responsive multifunctional PEGs.

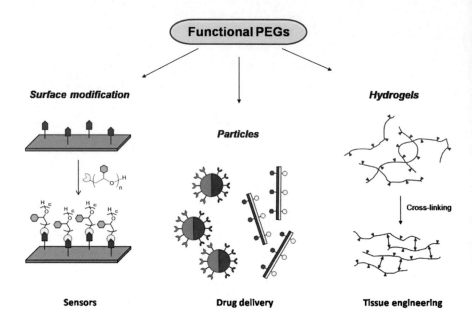

Scheme 15   Scope of multifunctional PEGs applications.

4   Results and Discussion

4.1   Anionic Ring-Opening Polymerization of EO and Its Derivatives

4.1.1   Overview

Polymerization of EO and its derivatives is possible both by anionic and cationic mechanisms.[132] The choice of the polymerization mechanism will mostly be defined by the presence of functional groups in the epoxide ring. Nevertheless often the preference is given to anionic ring-opening polymerization as it allows for better control over molecular weight.[96,132] Since living polymerization is free of termination processes and chains are still active even after complete consumption of monomer, synthesis of block copolymers by addition of a second monomer after full consumption of the first one or quantitative termination with functional agents are possible.

Anionic ring opening polymerization is typically performed in water free solvents and initiated by alkali-metal alkoxides.[61,96,132] Several polymerization parameters have to be considered in order to achieve efficient initiation and to maintain the control and living manner of polymerization over the course of reaction.

Water-Free Conditions

The biggest challenge in anionic polymerization of EO initiated by functional alkoxides is complete removal of water traces from the system, which not only cause termination reactions, but can also compete in an acid-base equilibrium with the initiating anion and lead to the formation of PEG-diol as a side product of initiation by hydroxyl anion.[96] It becomes a problem for further applications of thus obtained heterobifunctional PEGs, for example, drug loading as impurities of PEG-diol can cause undesired side reactions and may result in crosslinking of the loaded biomolecules.[133] Water-free environment can be achieved by the use of absolute solvents, proper initiating systems and performing the polymerization under the atmosphere of dry inert gases.

Solvents

Tetrahydrofuran (THF) is a commonly used solvent for the polymerization of EO. Initiating alkoxides and propagating PEG chains are fairly soluble in THF, for example, in comparison with toluene – another common solvent for anionic ring-opening polymerization. On the other hand, the complete removal of the solvent after polymerization is very easy as compared to dimethyl sulfoxide (DMSO) or diglyme and can be done with a common rotary evaporator. Finally, the solvents must be absolute to avoid initiation by hydroxyl anion (above). Absolute THF can be easily prepared by distillation from metal sodium under an inert gas atmosphere.

Temperature and Pressure

To increase the efficiency of initiation, polymerization of EO is often performed at high temperatures and pressures, which facilitate side reactions caused by the presence of any impurities traces in the reaction mixture or can even lead to the decomposition of a functional group introduced with the initiator.[62] Therefore the selection of efficient initiating system allowing for polymerization under mild conditions is of particular importance.

Counter Ion

Counter ion is one of the initiating system constituents, which contributes not only to the efficiency of initiation, but of the whole polymerization process. It has been shown, that the polymerization rate is highly dependent on the metal counter ion of the initiator and increases in the following sequence: $Li^+ < Na^+ < K^+ < Cs^+$.[96,134] In case of initiation by lithium compounds (butyl lithium, lithium naphthalene) no polymerization takes place because of the strong association between lithium and oxygen.[134,135] However, it has been recently shown that the addition of a phosphazene base complexes the lithium ion thus releasing the active alkoxide species and making polymerization possible.[136] In another example, polymerization of EO initiated by living polystyryllithium and polyisoprenyllithium was activated by the addition of triisobutylaluminum, thus allowing for the synthesis of diblock copolymers of EO with styrene and isoprene.[137] Although initiators with Na^+ counter ion are known to be used for the EO polymerization, they show much slower polymerization rates as well as low solubility in organic solvents as compared to K^+ and Cs^+ ones. Still the latter is less commonly utilized because of its high cost.[96] Polymerization by Na^+ and K^+ initiators can also be facilitated by addition of crown ethers.[138,139]

Initiating System

Different initiating systems relying on commercially available alkoxides such as *t*-BuOK,[110,140] or *in situ* prepared initiators from an alcohol and a deprotonating agent (Na,[138,141], K,[142] NaH,[65,143] *t*-BuOK,[144] diphenylmethylpotassium (DPMK),[64] KHMDS,[145] potassium naphthalenide (PN),[62,97,146] CsOH[77]) are known from literature. Often commercial or prepared initiating alkoxides require thorough removal of alcohol excess and/or water traces by boiling with benzene or toluene for some hours followed by azeotropic distillation in vacuum.[77,138] In case of *in situ* alkoxide formation via irreversible reaction of an alcohol with a base, which is much stronger than the initiating alkoxide, drying of initiator in vacuum is not necessary, since the traces of protic impurities would react with an excess of base. Therefore such systems simplify the initiating alkoxide preparation process and do not involve high temperatures which might cause degradation of the functional groups present in alkoxide.

Considering the discussed above factors, which influence the polymerization efficiency, for the following study on the EO polymerization the initiating system with 2-methoxyethanol and KHMDS as a base for the alkoxide formation was chosen. KHMDS is commercially available compound and is easy in use as there is no need for additional preparation before the polymerization as in case of DPMK[64] or PN.[62] The rationale for using 2-methoxyethanol[98] as the initiating alcohol, is as follows:

- methoxy group is sufficiently inert and stable under conditions of the polymerization,
- due to the presence of the methoxy group analysis of PEG's molecular weight by means of ^{1}H NMR spectroscopy is possible,
- the solubility of 2-methoxyethanolates organic solvents is higher than the solubility of methoxides.

4.1.2 EO Polymerization

With the chosen initiating system polymerization of EO was performed in THF, and different experimental parameters were tested in order to optimize the reaction. The obtained polymers were characterized by ^{1}H NMR and MALDI-TOF spectrometry (Table 1).

Table 1   Polymerization of EO[a].

entry	base	EO addition	t (°C)	EO[b] conv. (%)	$\overline{DP}_{n(NMR)}$[c,d]	$DP_{n(MALDI)}$[c,e]		
						polymer 1	PEG-diol	other side polymers
1	KHMDS	dist.	rt	<2	–	–	–	–
2	KHMDS	dist.	60	26	17	~21	~32	~30
3	KHMDS	cannula	rt	114[f]	118[f]	~111[f]	~28	–
4	PN	cannula	rt	93	108	~97	–	–

[a]For all polymerizations the feed ratio of initiator to monomer was 1:100, the reaction time was 72h. [b]EO conversions were determined gravimetrically. [c]Degree of polymerization. [d]Calculated based on ^{1}H NMR spectra. [e]Estimated based on MALDI-TOF-MS spectra. [f]Overrated conversion and $\overline{DP}_n$ values are due to the experimental inaccuracy of measuring the EO amount introduced into reaction.

For this study it was important to estimate the efficiency of the initiating system, since the next step would be synthesis of α,ω-heterotelechelic PEGs (Chapter 4.1.3). For this purpose MALDI-TOF-MS is the best available technique as it gives the exact values of molecular masses, thus it can be determined if the initiation was due to the introduced alkoxide or water, and if any side or decomposition reactions took place. It should be noted though that MALDI-TOF spectrometry doesn't allow for the quantitative determination of the relative amounts of different polymer chains present in the sample. Approximate $\overline{DP}_n$ values were estimated at maximums of molecular weights distributions.

In the first experiment, EO was distilled to the frozen initiating solution and polymerization was performed at room temperature. This resulted in a very low conversion of EO, just traces of the polymer were observed (Table 1, Entry 1).

Increasing the polymerization temperature up to 60°C improved the EO conversion (Table 1, Entry 2). Still the molecular weight of polymer **1** was much lower, than that calculated from the initiator to monomer feed ratio. This can likely be explained by the long time required for the transfer of EO into the reaction vessel and high possibility of leaking of gaseous EO during the transfer. Moreover, polymerization was accompanied by the formation of the PEG-diol and some other side polymers (Figure 1A), which should be attributed to the insufficient removal of water and other protic impurities from the initiating system.

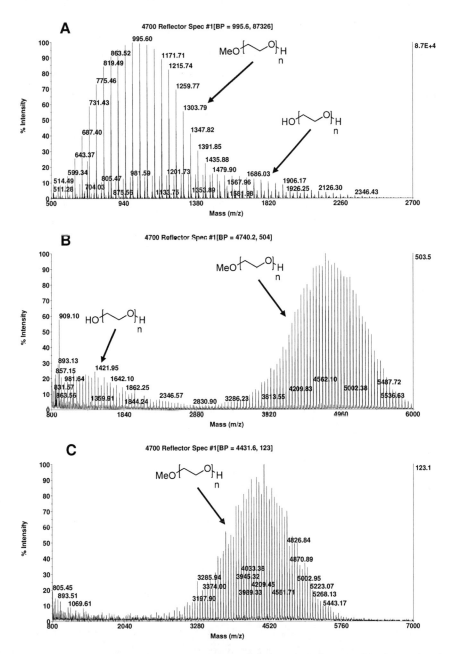

Figure 1 Comparable MALDI-TOF-MS spectra of polymer **1** obtained in the conditions corresponding to Table 1, Entry 2 (A), Entry 3 (B) and Entry 4 (C).

To further improve the process, the method of EO transfer to initiator was changed. EO was cooled to 0°C on the ice bath and was quickly transferred into the reaction vessel via cannula. The efficiency of the polymerization was improved dramatically even when it was performed at room temperature (Table 1, Entry 3). Overrated conversion is probably due to the difficulties in precise control of the EO amount introduced into the reaction. Still some traces of PEG-diol were present in the reaction mixture, although to a smaller extend (Figure 1B). This confirms the assumption that the reason for side polymers formation lies in low efficiency of the initiating system.

To verify this assumption, the base for the formation of alkoxide was changed for PN. Although PN has to be freshly prepared prior to polymerization, it was shown to be highly efficient and to allow for polymerization under relatively mild conditions (room temperature).[62] It is not only easy to control the conversion of an alcohol into the alkoxide due to the color change, but the tiny excess of PN also removes all traces of water or other protic impurities which might be present in the reaction mixture and lead to side reactions. Indeed, with the use of PN, not only EO conversion remained high, but also the obtained polymer was free of PEG-diol or any other impurities (Table 1, Entry 4; Figure 1C).

Due to its efficiency this initiating system can be used for the synthesis of α,ω-heterobifunctional PEGs.

4.1.3   Synthesis of α,ω-Heterotelechelic PEGs

Synthetic Approaches to α,ω-Heterobifunctional PEGs

As it has been discussed in the introduction part (Chapter 2.2.1), α,ω-heterobifunctional PEGs can be synthesized either via modification of hydroxyl groups of commercially available PEG or via EO polymerization with the use of suitable initiators and terminating agents.

The disadvantage of the first approach is difficulty in obtaining pure heterobifunctional PEGs, since differentiation of two identical hydroxyls of PEG mostly results in a mixture of products (not functionalized, monofunctionalized and bifunctionalized PEG-diol). Furthermore, the following modification reactions do not always proceed with quantitative yields (except for the click reactions) increasing the number of different bifunctional PEGs present in the product mixture. The similarity of physical properties of PEGs differed by just one or two functional groups makes their separation by crystallization impossible. One feasible solution is purification by column chromatography. It has been shown, that commercially available mPEG can be separated from impurities of PEG-diol chromatographically on silica gel.[82] However chromatographic purification, which is generally a time consuming technique, in case of separation of different bifunctional PEGs, has quite low efficiency and results in low yields because of the very close R_f values of these polymers.

On the other hand, initiation and termination of EO polymerization by appropriate agents bearing functional groups should result exclusively in the formation of the desired heterobifunctional polymer, assuming that the system is free of any protic impurities, which might interfere with the polymerization process. As it has been shown in Chapter 4.1.2, polymerization of EO initiated by alkoxide *in situ* formed from 2-methoxyethanol and PN satisfies the latter condition. Moreover, mild conditions of the polymerization should be compatible with a wider range of functional groups introduced with the initiator.

Based on these considerations the synthesis of α,ω-heterobifunctional PEGs was performed via the second approach.

EO Polymerization Initiated by Functional Alkoxides

EO polymerization initiated by functional alkoxides was performed in the conditions established for the synthesis of mPEG (Chapter 4.1.2) and the obtained polymers after recrystallization were characterized by ¹H NMR and MALDI-TOF spectrometry (Table 2).

The initiators used for this study are shown in Table 2. The goal was to introduce chemistries which can further be involved in so called "click" reaction.[122,147]

Table 2 α,ω-Heterobifunctional PEGs by anionic ring-opening polymerization of EO using functional initiators.[a]

entry	functional group	I:M feed ratio[b]	t_{add}[c] (°C)	EO conv.[d] (%)	$\overline{DP}_{n(NMR)}$[e,f]	$DP_{n(MALDI)}$[e,g]		
						polymer **2**	PEG-diol	other side polymers
1	**a**	1:200	rt	93	127	~87	–	~50
2	**b**	1:100	rt	100	no aromatic signals	–	~39	–
3			–78	100		–	~39	–
4	**c**	1:100	rt	26	44	~23	–	~15
5			–78	65	140	~59	–	~35
6	**d**	1:100	rt	100	150	~140	~35	–
7			–78	100	111	~100	–	–
8	**e**	1:100	rt	100	–	mixture of polymers		
9			–78	100	–			
10	**f**	1:100	rt	–	no polymerization			

[a]All polymerizations were performed at rt for 72 h. [b]The feed ratio of initiator to monomer. [c]Temperature, at which PN was added to the initiating alcohol for the alkoxide formation. [d]EO conversions were determined gravimetrically. [e]Degree of polymerization. [f]Calculated based on ^{1}H NMR spectra. [g]Estimated based on MALDI-TOF-MS spectra.

MALDI-TOF-MS spectra of polymer **3a** showed the presence of two series of molecular masses (Table 2, Entry 1; Figure 2). The masses corresponding to the peaks of the major series match the polymer with the following structure: HO-PEG-SH. This is however, in accordance with previously reported data showing that reduction of disulfides upon ionization from different matrixes such as α-cyano-4-hydroxy cinnamic acid commonly takes place.[148] The presence of a triplet at 2.74 ppm corresponding to the signal of CH_2S and absence of thiol proton in ^{1}H NMR spectrum of polymer **3a** also confirms the disulfide structure of the polymer. Unfortunately, assignment of any polymer structure to the minor series of molecular masses was not possible.

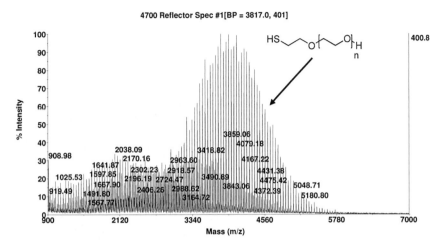

Figure 2   MALDI-TOF-MS spectrum of polymer **3a** obtained in the conditions corresponding to Table 2, Entry 1.

When 9-anthracenemethanol (**2b**) was used as initiating alcohol for EO polymerization, no polymer **3b** was found in MALDI-TOF-MS spectra of the product. Apparently, the only polymer detected by MALDI-TOF spectrometry was PEG-diol (Table 2, Entry 2). ^{1}H NMR spectra of the recrystallized product also revealed no presence of aromatic signals. The only signals detected by ^{1}H NMR spectroscopy were those of PEG backbone. This fact can be explained by hydrogenolysis of anthracenemethanol during the alkoxide formation, similarly to that previously observed during benzyl alcohols reduction by dissolving metals.[149] Furthermore, in ^{1}H NMR spectra of crude polymer no benzyl protons were detected, instead a new singlet appeared at 3.1 ppm, and was assigned to the methyl group of methylanthracene, which also confirms, that hydrogenolysis takes place. To suppress undesired hydrogenolysis the addition of PN to 9-anthracene methanol was performed at –78°C (Table 2, Entry 3). Unfortunately, this precaution did not change the polymerization outcome and PEG-diol was detected as the only product of polymerization.

To overcome this problem, alcohol **2c** was synthesized and used in polymerization of EO (Table 2, Entry 4). This partially improved the results, and the polymer **3c** was obtained as a major product. However, the conversions were very low, and a side polymer with lower molecular weight was detected in the product mixture. Some side processes during alkoxide formation may still occur. When addition of PN to alcohol **2c** was performed at −78°C, the conversion of EO has been improved, however the side polymer still was detected in the product mixture (Table 2, Entry 5). It is difficult to speculate on the origins of the side reaction, but it is highly probable that C=O bond of amide moiety is involved.

Bigger success was achieved with alcohol **2d** bearing protected aldehyde group. Although hydrogenolysis was partially observed when alkoxide was formed at room temperature (Table 2, Entry 6), it was completely suppressed when PN addition was performed at −78°C (Table 2, Entry 7). Not only was quantitative conversion of EO achieved, but characterization of the polymer **3d** by ^{1}H NMR and MALDI-TOF spectrometry revealed no presence of side product (Figure 3).

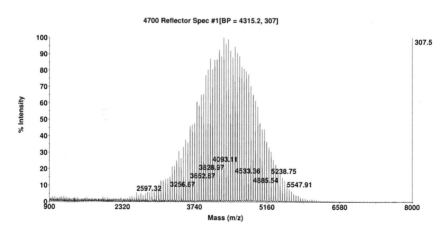

Figure 3   MALDI-TOF-MS spectrum of polymer **3d** obtained in the conditions corresponding to Table 2, Entry 7.

Polymerization initiated by 2-azidoethanolate resulted in a complicated mixture of polymers (Table 2, Entry 8). There was no change in the polymerization outcome when PN addition to the alcohol was performed at −78°C (Table 2, Entry 9).

Finally, no polymerization occurred when alcohol with protected maleimide group **2f** was used (Table 2, Entry 10). Immediate change of the color of alcohol solution upon addition of PN to dark violet provides evidence of the reaction of alcohol with PN. Even with protection the molecule is still quite reactive and can not be used for EO polymerization.

The above study showed that although efficient initiation by can be achieved with use of alcohol/PN system, the range of functional groups introduced with initiator is limited. In several cases optimization of conditions can diminish side processes. However some groups can not be introduced via anionic polymerization. For such cases coordination polymerization can be a suitable solution.[61]

The following study is concentrated on the synthesis of multifunctional polymers. In order to simplify the results interpretation and prevent misunderstanding of the nature of side processes the initiating alkoxides further used do not bear reactive groups. Although the polymerization of substituted epoxides combined with initiation by functional alkoxides is of course possible and can be used as an additional source of orthogonal reactive groups.

4.1.4   Synthesis of Homomultifunctional PEGs

4.1.4.1   Synthesis of Functional Epoxides

Synthesis of functional epoxides commonly relies on the following strategies:

-   reactions of ECH (**4**)
-   reactions of glycidol (**5**)
-   epoxidation of alkenes

ECH and glycidol are commercially available epoxides both in racemic and enantiomerically pure forms. Thus many approaches for the synthesis of functional epoxides are based on their modification.

A series of functional epoxides has been prepared by utilizing the mentioned above routes (Scheme 16). The following description of the synthesis is divided into blocks according to the used strategy.

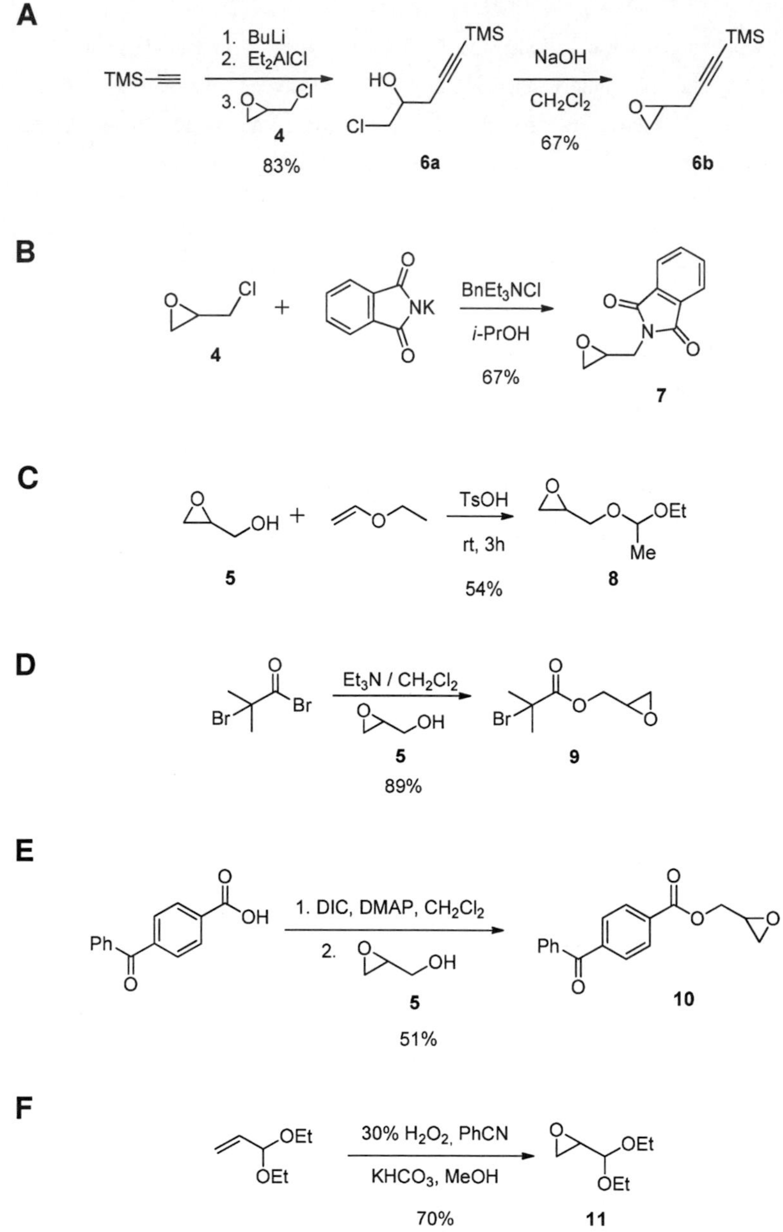

A

B

C

D

E

F

Scheme 16   Synthesis of functional epoxides.

Reactions of ECH

Reactions involving ECH proceed via opening of the epoxide ring by a functional nucleophile, resulting in the formation of the corresponding chlorohydrine, which in many cases can be isolated and characterized. Following intramolecular nucleophilic substitution of chlorine leads to the ring closing yielding the target epoxide.[96,150,151] This procedure was applied for the synthesis of epoxide **6** with TMS-protected alkyne, which was obtained with the overall yield after two steps 55% (Scheme 16A). TMS protection can be easily cleaved upon treating with KF in MeOH.[152,153]

With the use of phase transfer catalysts both steps can be done in one pot without isolation of the intermediate chlorohydrine.[96,154] Thus epoxide **7** with protected amino group was synthesized (Scheme 16B). Removal of phthaleimide group is typically performed by hydrazine hydrate in ethanol, although other reagents are known to be used.[155]

Reactions of Glycidol

Synthesis of linear PG is performed by polymerization of differently protected glycidols.[99] Most commonly acetal protection is used due to the ease of cleavage by acid treatment. Epoxide **8** was prepared by acid catalyzed reaction of glycidol with methyl vinyl ether (Scheme 16C) according to the previously reported procedure.[156]

Approaches to synthesize other functional epoxides starting from glycidol typically involve esterification reactions.[157] Thus epoxide **9** with an ATRP initiating group (Scheme 16D) and epoxide **10** with benzophenone group (Scheme 16E) were synthesized. The latter can be used for photoinitiated crosslinking.

Epoxidation of Alkenes

Many methods relying on alkene's epoxidation, including stereospecific epoxidation with the use of chiral agents have been developed.[158] Epoxidation via electrophilic mechanism is common for electron rich alkenes and most typically uses *meta*-chloroperoxybenzoic acid (*m*CPBA) as epoxidating agent. The nucleophilic mechanism is more suitable for electron deficient alkenes and usually performed with the use of hydrogen peroxide.

GADEA (**11**)[159] was prepared via epoxidation of corresponding alkene by hydrogen peroxide in the presence of benzonitrile (Scheme 16F). Hydrogen peroxide reacts with benzophenone forming benzenecarbimideperoxoic acid. The latter acts as an oxidizing agent converting alkene into epoxide according to electrophilic mechanism.

Although not all synthesized epoxides may be suitable for polymerization (Chapter 4.1.4.2.3) because of the instability of functional groups under basic conditions of polymerization, there is still a large field for their applications based on the reactivity of epoxides toward nucleophiles. They can be used as terminating agents in anionic polymerization[160] or serve as short linkers for surface modification (Chapter 4.2).

4.1.4.2 Polymerization of Functional Epoxides

4.1.4.2.1 Overview

Compared to polymerization of EO, polymerization of functional epoxides is connected with several further issues:

- presence of substituent(s) can result in different kinetics of polymerization,
- functional group(s) can be sensitive to polymerization conditions,
- termination reactions due to the abstraction of methylene proton from monomer or polymer are possible

Epoxides Reactivity

The decreased reactivity of functional epoxides in anionic ring-opening polymerization as compared to EO can be expected because of the steric hindrance of the epoxy ring created by the presence of a substituent. This assumption has been confirmed by a study by Heatle et al. on the comparative reactivity ratios of EO and PO.[161] The reported results showed that EO is more reactive toward the propagating chain than PO, and thus their copolymers have a gradient composition. Later Frey and co-workers studied kinetics of EO copolymerization with different glycidyl ethers (EEGE[77], AGE,[98] DBAG[119], IGG,[117] EVGE[101]) by monitoring the consumption of comonomers at different polymerization times in ^{1}H NMR spectra. The microstructure of the copolymers with various EO:glycidyl ether molar ratios was analyzed by qualitative comparison of triad intensities in ^{13}C NMR spectra. In all cases the authors concluded that copolymerization results in a random comonomer distribution. However recently Lee at al. showed that actual reactivities of AGE and EVGE are higher than that of EO.[162] They developed a new approach for the calculation of the comonomer's reactivity ratios based on the qualitative comparison of resonances in ^{1}H NMR spectra, which are sensitive to the identities of the first two monomers added to the initiator.[162] Moreover, the reactivity of EVGE was found to be higher than that of AGE. To explain these quite unexpected results the authors proposed a polymerization mechanism via coordination of the epoxide to the potassium counterion, which was confirmed by transition-state density functional calculations. Still, for the epoxides with more bulky substituents and which are not glycidyl ethers (thus there is no additional coordination of glycidyl ether oxygen to the counterion) polymerization rates can considerably decrease as compared to EO.

Functional Group Side Reactions

Not all functional/protected functional groups can be stable in the highly basic conditions of anionic polymerization. However, sometimes changing the polymerization conditions, such as lowering temperature and pressure or use of different solvents compared to the polymerization of EO, can minimize the side processes and lead to pure multifunctional polymer. For example, isomerization of allyl group during the polymerization of AGE can be suppressed at temperatures below 40°C.[97]

Termination Reactions

The biggest problem connected with polymerization of substituted epoxides is termination reactions due to the abstraction of methylene proton from monomer or polymer (Scheme 17).[94,144] This process limits the molecular weights and increases the polydispersities of the synthesized polymers.

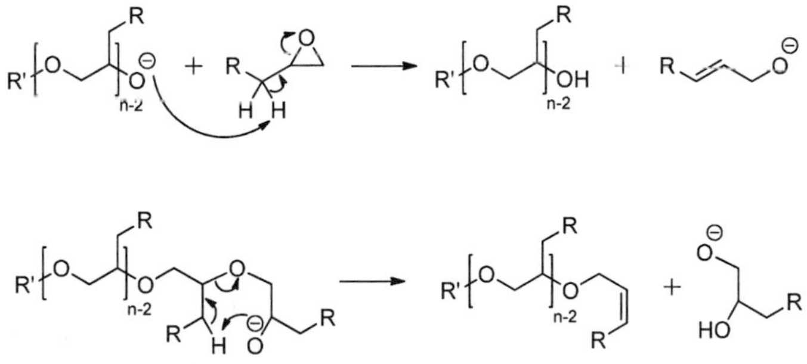

Scheme 17 Termination reactions during anionic polymerization of functional epoxides.[94,144]

Many studies have been performed on the polymerization conditions of PO and phenyl glycidyl ether (PhGE), where termination reactions have been observed on particularly large extension.[92,93,163] The side processes were partially suppressed at low temperatures and with the use of alcoholate/alcohol initiating mixture and softer counter ions or in the presence of crown ethers.[94] As opposed to traditional alkoxide initiated anionic polymerization, PPOs with molecular weight up to several thousand with negligible unsaturation were prepared with the use of coordination catalysts, such as double metal cyanides.[164,165] For the higher analogues of PPO, the terminating processes are less dramatic. Thus poly(butylene oxide) (PBO), poly(hexylene oxide) (PHO) and poly(octylene oxide) (POO) with molecular weights in the range 50 000 – 100 000 and polydispersity indexes (PDI) < 1.1 were obtained at low temperature with the use of crown ether.[94] Hans et al. performed a similar investigation of the EEGE polymerization.[144] The authors concluded that the molecular weight of PEEGE obtained by conventional anionic polymerization is limited to 30 000 because of the termination. However, the authors did not go below 60°C with potassium alkoxide as an initiator. It was also suggested that the approach toward synthesis of high molecular weight PEEGEs with low PDIs should rely on the decrease of the propagating chain basicity. This can be achieved by means of coordination anionic polymerization, for example when partially hydrolyzed diethylzinc was used as a catalyst. Thus synthesis of high molecular weight PEEGE was possible.[166,167] The drawback of this method is high polydispersities of the obtained polymers. Another approach relies on the use of triisobutylalluminium.[96,144,168] These studies showed that termination is greatly dependent on the polymerization conditions, particularly on the reaction temperature and the initiating system. Thus "softer" K^+ or Cs^+ initiators as well as temperatures close to ambient minimize the termination reaction and lead to a polymer with a controlled structure.

Furthermore, the presence of functional groups having asymmetric atoms leads to the formation of different diastereomers, which causes broadening of the signals in NMR spectra and complicates the picks assignment.

Since polymerization of substituted epoxides gives several additional limitations in comparison to EO polymerization, new study on the polymerization conditions is required.

Specificity of AGE Polymerization

AGE is one of the few known EO derivatives suitable for direct anionic ring-opening polymerization (Chapter 2.2.3). The interest in PAGE results from the potential to further modify the polymer via thiol-ene reaction, which allows for quantitative conversions of PAGE into various new multifunctional polymers, which could not be obtained directly via polymerization of the corresponding functional epoxide. Besides, AGE is readily available commercially. However, there are two additional issues arising from the presence of allyl groups.

First, additional termination can be caused by the abstraction of a proton from the allyl position. This fact was mentioned in the study done by Erberich et al., who concluded that the polymerization of AGE can be controlled just up to ca. 80% conversions.[99] They suggested that at higher conversions termination by abstraction of allyl proton takes place and compete with propagation. Later Lee et al. performed a thorough investigation of the AGE polymerization with PN used for the formation of initiating alkoxide.[97] They observed no transfer or termination reaction, suggesting that the increase of PDIs of PAGE with high molecular masses are due to the chain coupling or radical coupling of backbone allyl substituents.

Second, isomerization of allyl groups is possible. In organic chemistry, the ally group is used as a protecting group for alcohols.[169] The conditions for the cleavage of allyl protection are similar to the conditions of epoxides anionic polymerization (Scheme 18).[98] This is quite unfortunate as isomerized allyl group can easily undergo hydrolysis resulting in the presence of hydroxyl groups within the polymer chain, what can considerably change the properties of the polymer and provide possibilities for further modification. However, it was shown, that lowering the polymerization temperature from 60 to 40°C resulted in a significant decrease of the isomerization product.[98] The copolymers P(EO-*co*-AGE) were obtained with isomerization degree 0–10%. Lee et al. also underlined the strong dependence of the amount of isomerization product in PAGE on temperature.[97] According to the reported results, the isomerization is negligible when polymerization of AGE is performed below 40°C independently of other conditions (in solution or bulk). The authors used PN for the formation of initiating alkoxide. However, in this study no comparative investigation of different initiating systems was carried out. Still, the initiating system is one the most important factors influencing the outcome of the polymerization (Chapter 4.1.1).

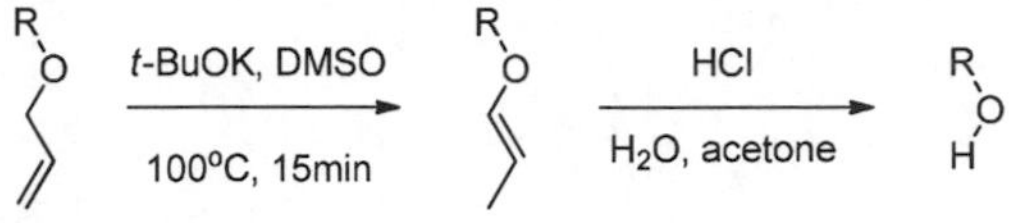

Scheme 18   Cleavage of allyl group into corresponding alcohol by base-catalyzed isomerization followed by acid hydrolysis.[98]

Here a comparison of different initiating systems that are known from literature to be used for polymerization of substituted epoxides is presented.

4.1.4.2.2 AGE Polymerization

First, estimation of the efficiency of different initiating systems by comparison of monomer conversions and amount of side products was done. In this part of the study isomerization was not taken into consideration and polymerization was performed at 60°C since previously for the polymerization of EEGE it was shown that termination reactions are minimized at this temperature.[144] Moreover, it was shown that the isomerization does not interfere with propagation and does not lead to any termination process.[97]

The following initiating systems were further tested: t-BuOK, alcohol + BuLi +crown, alcohol + NaH, alcohol + t-BuOK, alcohol + KHMDS, alcohol + PN. Several other reaction parameters such as solvent and monomer/initiator ratio were also varied. The results are summarized in Table 3.

Table 3   Optimization of the conditions for AGE polymerization.[a]

1) base

ROH $\xrightarrow{\quad\quad}$ $RO{\left[\!\!\begin{array}{c}O\\\\O\end{array}\!\!\right]}_n H$ + $HO{\left[\!\!\begin{array}{c}O\\\\O\end{array}\!\!\right]}_n H$

2) n (epoxide)

12

a R = t-Bu
b R = MeOCH$_2$CH$_2$

entry	base	solvent	I:M feed ratio[b]	t (°C)	^{1}H NMR[c]			DP$_{n(MALDI)}$[d,e]		
					AGE conv. (%)	$\overline{DP}_n$[e]		polymer **13**	PAGE-diol	other side polymers
1	t-BuOK	THF	1:10	60	89	15		~8	~7	–
2	t-BuOK	THF	1:25	60	40	–		–	–	–
3	t-BuOK	DMSO	1:25	60	99	24		~20	~7	~7
4	n-BuLi[f] 12-c-4	THF	1:50	60	no polymerization					
5	NaH[f]	THF	1:25	60	79	–		~20	~7	–
6	t-BuOK[f]	THF	1:25	60	99	59		~31	~7	–
7	KHMDS[f]	THF	1:50	60	99	44		~27	~6	~8
8	KHMDS[f]	DMSO	1:50	60	99	44		~23	~7	~7
9	PN[f]	THF	1:30	60	80	22		~18	–	–
10	PN[f]	THF	1:30	rt	79	24		~ 20	–	–

[a]All polymerizations were performed for 48 h. [b]The feed ratio of initiator to monomer. [c]Calculated based on ^{1}H NMR spectra. [d]Estimated based on MALDI-TOF-MS spectra. [e]Degree of polymerization. [f]Together with 2-methoxyethanol.

Polymerization initiated by t-BuOK, when the feed ratio of initiator to AGE was 1:10, resulted in a mixture of polymer **12a** and PAGE-diol (Table 3, Entry 1). Presence of PAGE-diol is probably due to the difficulties in removing all water traces from the initiator. Occurrence of PAGE-diol in the product explains the overrated degree of polymerization ($\overline{DP}_n$) values estimated by ^{1}H NMR.

Increasing the number of monomer equivalents, resulted in very low conversion (Table 3, Entry 2), which can be explained by low solubility of the initiator in THF together with slow initiation by tertiary alkoxide. Because of the low conversion, the further study of the product content by MALDI-TOF spectrometry was not performed. To check if the solubility of initiator was indeed a problem the solvent was changed to DMSO (Table 3, Entry 3). Thus excellent conversions were achieved; however PAGE-diol as well as other polymer impurities were still found in the reaction mixture in a considerable amount. This leads to conclusion that this initiating system was not efficient enough.

Further experiments involved *in situ* formation of initiating alkoxide from 2-methoxyethanol and different bases. Lithium alkoxide formed upon treating of 2-methoxyethanol with n-BuLi was inefficient despite of the presence of crown ether (Table 3, Entry 4). NaH showed no further improvement compared to the results obtained with *t*-BuOK (Table 3, Entry 5). Next, *t*-BuOK was used as deprotonating agent. To shift the acid-base equilibrium toward 2-methoxyethanolate formation a big excess of alcohol was used, which together with formed *t*-BuOH was subsequently removed *in vacuo*. The results were similar to those obtained with other initiators: though good conversion of AGE was achieved, the product contained both target polymer **12b** and PAGE-diol (Table 3, Entry 6).

The drawback of the initiating systems described thus far is insufficient removal of water and other impurities traces, which led to the formation of PAGE-diol and other side polymers. Therefore further experiment involved the use of a stronger base (KHMDS or PN) to form the alkoxide from 2-methoxyethanol. Unexpectedly, despite quantitative conversions of the monomer, the use of KHMDS did not result in efficient removal of water traces and PAGE-diol was still detected in the product (Table 3, Entries 7,8; Figure 4). Different solvents were not shown to influence the obtained results.

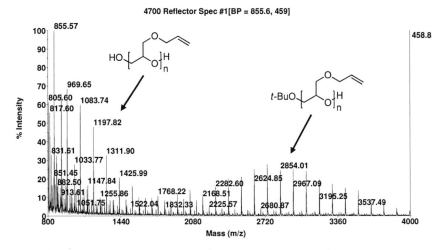

Figure 4   MALDI-TOF-MS spectra of polymer **12b** obtained in the conditions corresponding to Table 3, Entry 7.

Compared to KHMDS, PN gives an additional advantage of visual control of the alkoxide formation, due to the color change. Solution of PN in THF has a characteristic deep green color, which disappears upon reaction with protic compounds. Addition of PN to 2-methoxyethanol was performed till the green color remained, meaning complete conversion of the latter into the corresponding alkoxide. Moreover, any traces of water, which might be present in the initiator solution would thus be completely removed. Indeed, this initiating systems lead to the formation of the target polymer **12b** only without any traces of side reactions (Table 3, Entry 9). Furthermore, the same efficiency of the polymerization was also achieved at room temperature (Table 3, Entry 10; Figure 5). This initiating system has been employed for the further study of the polymerization of AGE.

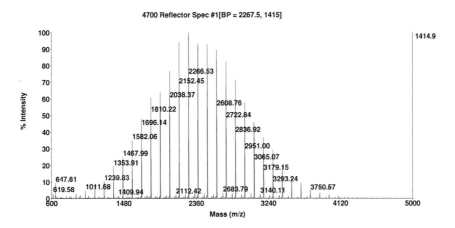

Figure 5   MALDI-TOF-MS spectra of polymer **12b** obtained in the conditions corresponding to Table 3, Entry 10.

To see if further improvement of the yields is possible, the polymerization of AGE was performed in bulk at 30°C. Polymerization in bulk should proceed with higher yields due to the smaller distances between molecules and higher possibilities for reactive species to come in contact. This should in turn result in a shorter polymerization time. A slight increase in temperature should also facilitate the polymerization. However this temperature would still be below 40°C, which should minimize isomerization.[97] To have a better picture of the processes taking place when polymers with higher molecular weights are synthesized the number of monomer equivalents to was increased up to 60. The results of bulk polymerization are shown in Table 4, Entry 1. For comparison the results of polymerization in solution are also included (Table 4, Entry 2). As expected bulk polymerization proceeds with quantitative yields within 24 h, while polymerization in solution resulted in 96% conversion after 72 h. The percentage of isomerized allyl groups was very low in both cases; however when the polymerization was performed in bulk isomerization was hardly detectable. This is probably due to the fact that polymerization in solution was done for a longer time period.

Table 4   Polymerization of AGE in bulk and in solution[a].

entry	solvent	t (°C)	polymeriz. time (h)	^{1}H NMR[b]			GPC[c]		
				AGE conv. (%)	d_n	isomer.[e] (%)	M_n (g/mol)	M_w (g/mol)	PDI[f]
1	–	30	24	99	64	0.6	7200	8200	1.13
2	THF[g]	rt	72	96	60	1.8	7300	7800	1.08

[a]For all polymerizations the feed ratio of initiator to monomer was 1:60. [b]Calculated based on ^{1}H NMR spectra. [c]Determined by GPC analysis in THF relative to PS standards. [d]Degree of polymerization. [e]Percentage of isomerized allyl groups in polymer. [f]Polydispersity index = M_w/M_n. [g]25 w/v% solution.

Next, both polymers (obtained in solution or in bulk, Table 4) were analyzed by GPC. On the GPC profiles of the obtained polymers, a shoulder in a higher molecular masses range is apparent. The molecular weight of this shoulder is approximately two times higher than that of the target polymer (Figure 6). Importantly, the size of the shoulder was much bigger when the polymerization was done in bulk compared to the polymerization in solution.

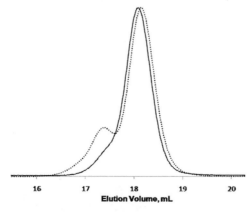

Figure 6   Comparative GPC chromatograms of polymer **12b** synthesized in the conditions corresponding to Table 4, Entry 1 (dotted line) and Entry 2 (solid line).

To see how the GPC profile of polymer **12b** changes with the molecular weight of the polymer, PAGEs with different feed ratios of initiator to AGE in the conditions corresponding to those in Table 4, Entry 1 were synthesized. The results are summarized in Table 5.

Table 5   Characterization of polymers **12b** with different molecular weights synthesized by bulk polymerization of AGE.[a]

entry	I:M feed ratio[b]	^{1}H NMR[c]		GPC[d]		
		AGE conv. (%)	$\overline{DP}_n$ [e]	M_n (g/mol)	M_w (g/mol)	PDI[f]
1	1:30	99	31	4100	4500	1.11
2	1:60	99	64	7200	8200	1.13
3	1:100	99	98	11300	12800	1.14
4	1:200	89	172	14700	19600	1.34

[a]Conditions of polymerization were the same as in Table 4, Entry1. [b]The feed ratio of initiator to monomer. [c]Calculated based on ^{1}H NMR spectra. [d]Determined by GPC in THF relative to PS standards. [e]Degree of polymerization. [f]Polydispersity index = M_w/M_n.

As seen in Figure 7, the shoulder on GPC profiles of the synthesized polymers grows with the increase of molecular weight of the polymer. This can be explained by cross-coupling of allyl moieties, most possibly via a radical mechanism. The assumption that the coupling proceeds via a radical mechanism and does not involve reactive ends of the propagating polymers is confirmed by the appearance of the second shoulder in case of polymers **12b** with high $\overline{DP}_n$ (Figure 7, solid line). The molecular weight of the second shoulder corresponds to approximately three molecular weights of the target polymer. Decrease of the shoulder in case of polymerization in solution (Figure 6, solid line) also supports the hypothesis of chain coupling, as in solution the chains are solvated and not so close to each other, that is why their coupling is less probable. However polymerization in solution resulted in lower conversions and longer reaction time, which in turn led to higher isomerization degree. Assuming that the coupling is of radical origin it should be possible to diminish it in radical free environment. Therefore further experiments with an emphasis on reactant preparation including thorough absolution and degassing of AGE were performed.

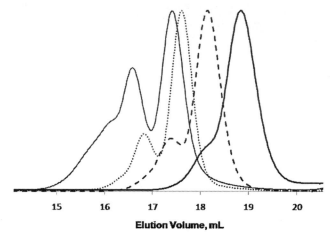

Figure 7   Comparative GPC chromatograms of polymer **12b** with different molecular
weights synthesized in the conditions corresponding to Table 5, Entry 1 (bold
solid line), Entry 2 (dashed line), Entry 3 (dotted line) and Entry 4 (solid line).

Previously, the preparation of AGE for polymerization included stirring with CaH_2 under Ar

atmosphere at 60°C for 2–3 h followed by distillation prior to polymerization (Table 6, Entry

1). This obviously did not remove the traces of oxygen, which might act as an initiator of

radical processes during the polymerization and cause coupling of polymeric chains. To avoid

traces of oxygen, a thorough degassing of AGE prior to polymerization was performed. Lee et

al. also reported previously on the degassing of AGE before polymerization together with

distillation of AGE from BuMgCl as a drying agent.[97] In this work, they managed to reach

the highest conversions of AGE accompanied by minimum side reactions, which must be at

least partially due to the monomer preparation technique. For the current study the initiator to

monomer feed ratio was chosen to be 1:100, because, as it was shown before (Table 5), full

conversion of monomer could be achieved within 24 h, and a shoulder peak can be clearly

distinguished in GPC profile. First, AGE was distilled from CaH_2 under Ar atmosphere. Next

the monomer was degassed via six freeze-pump-thaw cycles, followed by distillation from

BuMgCl. Polymerization of thus prepared AGE resulted in high conversion and narrow

molecular weight distribution (Table 6, Entry 4). The shoulder on the GPC profile of the

polymer **12b** decreased dramatically (Figure 8, dashed line). For comparison the results of

bulk and solution polymerization of AGE, prepared as before by just distillation from CaH_2,

are included in the Table 6 (Entry 1 and 2 correspondingly) and Figure 8 (dotted and bold

solid lines correspondingly). The advantage of getting low PDIs by solution polymerization is negated by very low conversion. Increase of conversion by longer reaction time is accompanied by the increase of PDI (Table 6, Entry 3). Therefore, optimization of results can only be achieved by the use of thoroughly prepared reactants completely free of oxygen and water.

Table 6   Conditions for polymerization of AGE and characterization of the obtained polymers **12b**.[a]

entry	AGE preparation	solvent	t (°C)	polymerization time (h)	AGE conv.[b] (%)	PDI[c]
1	- dist. from CaH$_2$	–	30	20	98	1.21
2	- -	THF	30	20	44	1.08
3	- -	THF	30	72	97	1.12
4	- dist. from CaH$_2$ - degassing - dist. from BuMgCl	–	30	20	96	1.09
5	- dist. from CaH$_2$ - dist. from BuMgCl - degassing	–	30	20	97	1.07
6	- -	–	30	12	91	1.10
7	- -	–	rt	20	94	1.10

[a]For all polymerizations the feed ratio of initiator to monomer was 1:100, polymerization time 20 h. [b]Calculated based on ^{1}H NMR spectra. [c]Polydispersity index = M_w/M_n, determined by GPC in THF relative to PS standards.

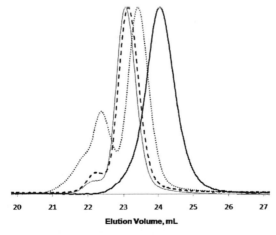

Elution Volume, mL

Figure 8   Comparative GPC chromatograms of polymer **12b** synthesized in the conditions corresponding to Table 6, Entry 1 (dotted line), Entry 2 (bold solid line), Entry 4 (dashed line) and Entry 5 (solid line).

Furthermore, the sequence of actions during AGE preparation was changed and first distillation from BuMgCl was performed followed by degasing. This change resulted in a slight further improvement (Table 6, Entry 5; Figure 8, solid line). Polymerization of thus prepared AGE for shorter time (Table 6, Entry 6) and at lower temperature (Table 6, Entry 7) resulted in lower conversions and led to no further improvement of PDIs.

Purification

PAGE is a viscous oil and can not be purified by standard crystallization as in case of bifunctional PEGs. Traces of unreacted monomer can be removed under high vacuum, however naphthalene and other solid impurities would still be present in the reaction mixture. Chromatographic purification on silica gel has been previously applied for the purification of bifunctional PEGs from the nonreacted PEG-diol.[82] Furthermore, Hruby et al. reported on the purification of mPEG-*b*-PAGE from PAGE homopolymer on silica gel with a chloroform/isopropyl alcohol mixture as eluent.[143].

Here purification of homomultifunctional PAGE from the low molecular weight impurities was also performed chromatographically. The purification procedure was changed compared to those previously reported and consisted of the following. Low molecular weight impurities including unreacted monomer and naphthalene were eluted from silica by chloroform. Pure polymer was subsequently successfully washed from the column by chloroform/methanol (10:0,7) mixture. ^{1}H NMR and MALDI-TOF-MS spectra as well as GPC analysis of thus purified PAGE showed that no degradation, hydrolysis or loss of high molecular weight fraction takes place (Figures 9, 10, 11).

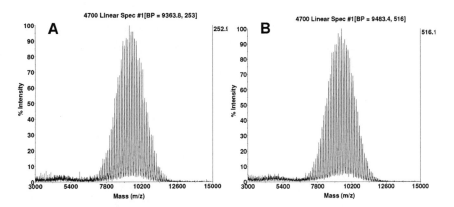

Figure 9   MALDI-TOF-MS spectra of **12b** obtained in the conditions corresponding to Table 6, Entry 5 before (A) and after (B) chromatographic purification on silica gel.

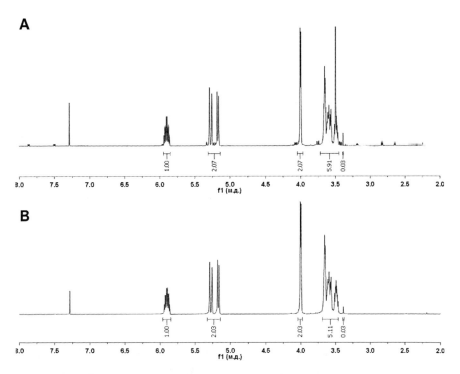

Figure 10 ^{1}H NMR spectra of **12b** obtained in the conditions corresponding to Table 6, Entry 5 before (A) and after (B) chromatographic purification on silica gel. Solvent CDCl$_3$.

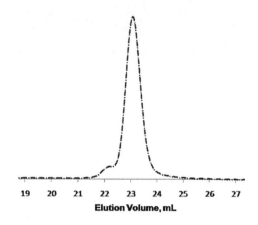

19   20   21   22   23   24   25   26   27
Elution Volume, mL

Figure 11   Gel   permeation   chromatograms   of   **12b**   obtained   in   the   conditions
corresponding to Table 6, Entry 5 before (dotted line) and after (dashed line)
chromatographic purification on silica gel.

4.1.4.2.3 Polymerization of Other Functional Epoxides

The polymerization of different substituted epoxides was carried out under the previously

optimized conditions with 2-methoxyethanol as initiator. The obtained polymers were

characterized by ^{1}H NMR and MALDI-TOF spectrometry as well as by GPC. The results are

summarized in Table 7.

Table 7   Polymerization of different functional epoxides.

entry	epoxide/ polymer	abbrev. (epoxide / polymer)	R	I:M feed ratio[a]	conv.[d] (%)	$\overline{DP}_n$[e]	M_n (g/mol)	M_w (g/mol)	PDI[f]
1[g]	13 / 14	BnGE/ PBnGE	(R: CH₂OCH₂Ph)	1:30	99	32	4000	4400	1.09
2[g]				1:60	98	59	7100	7600	1.06
3[g]	8 / 15	EEGE/ PEEGE	(R: CH₂O-CH(Me)-OEt)	1:30	91	28	3700	4000	1.08
4[g]				1:60	79	47	6100	6600	1.09
5[h]	11 / 16	GADEA/ PGADEA	(R: CH₂-CH(OEt)₂)	1:30	90	28	3500	3800	1.08
6[h]				1:60	83	51	5300	5600	1.06

[a]The feed ratio of initiator to monomer. [b]Calculated based on ^{1}H NMR spectra. [c]Determined by GPC in THF relative to PS standards. [d]Conversion of monomer. [e]Degree of polymerization. [f]Polydispersity index = M_w/M_n. [g]Polymerization was done in THF solution at rt for 72 h. [h]Polymerization was done in bulk at 30°C for 24 h.

In some cases polymerizations were still performed in THF solution (Table 7, Entries 1–4), because of the high viscosity of the reaction mixtures in the absence of solvent and impossibility for proper stirring during the reaction. This can be done for the epoxides other than AGE since no side processes caused by the presence of allyl groups (isomerization, radical cross-coupling) are possible. Thus, although reactions in solution proceed for a longer time than in bulk, it doesn't lead to undesired side reaction. Chain transfer reactions should also be limited because of low polymerization temperature (room temperature). Indeed, GPC traces of the obtained polymers were symmetric (Figure 12) and PDIs were low (Table 7).

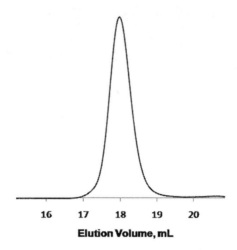

Figure 12   Gel permeation chromatogram of **14** obtained in the conditions corresponding
to Table 7, Entry 2.

As could be expected the reactivity of the epoxides decreases when more bulky substituent is
introduced. Thus in the same polymerization conditions benzyl glycidyl ether (BnGE) was
polymerized with the highest conversions (Table 7, Entries 1–2). Conversions of EEGE were
still high, but lower than those of BnGE (Table 7, Entries 3–4). Finally, when epoxide had a
secondary substituent, the conversions in polymerizations performed in solution were so low
(<20%), that the conditions had to be changed, and polymerization was done in bulk at 30°C
(Table 7, Entries 5–6).

However, not all functional groups are stable under strongly basic conditions of anionic
polymerization. In case of epoxides **6, 7, 9, 10** (Chapter 4.1.4.1) either no polymerization took
place or a complicated mixture of products was obtained. This is explained by the presence of
electrophilic groups (ester, ketone, imide), which interfere with the strongly nucleophilic
alkoxides leading to decomposition or suppression of the polymerization.

PBnGE and PEEGE are precursors of PG. Cleavage of benzyl group can be done by
hydrogenolysis,[115,116] while acetal protection is easily removed in acidic media.[77,99] The
possibility to remove protections orthogonally can be further used, for example, for sequential
modification of BnGE and EEGE copolymers, due to stepwise deprotection. Beside, BnGE
can be used to increase hydrophobic properties of a polymer when copolymerized with EO or
other functional epoxides, or represent a hydrophobic block when block copolymers with

irregular properties are desired. GADEA is an acetal protection for aldehyde group and can be cleaved upon treatment with acid.[73,74]

Purification of the obtained PEEGE and PGADEA should be taken with caution, since acetal groups are sensitive to acidic conditions, and cleavage of protecting groups can take place during chromatographic purification on silica gel. Nevertheless PGADEA turned out to be stable on silica gel: neither loss of acetal protecting groups was detected by ^{1}H NMR and MALDI-TOF-MS spectra (Figure 13A) nor decrease of molecular weight was shown by GPC (Figure 13B).

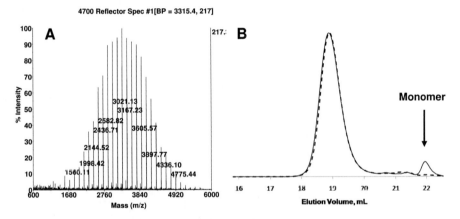

Figure 13   MALDI-TOF-MS spectrum (A) of purified polymer **16** obtained in the conditions corresponding to Table 7, Entry 5 (A) and gel permeation chromatograms of this polymer before (solid line) and after (dashed line) chromatographic purification on silica gel (B).

PEEGE however is less stable on silica gel and partial deprotection takes place. Nevertheless, the deprotection degree is very low and could not be detected by GPC. Partial cleavage of hydroxyl groups can be detected in ^{1}H NMR spectra by decrease of M_n and in MALDI-TOF-MS spectra by appearance of new molecular masses corresponding to partially deprotected polymer (Figure 14). Nevertheless the degree of deprotection is very low and doesn't change the retention factor (R_f) of the polymer, which allows purification of PEEGE without the loss of polymers. Column chromatographic purification of PEEGE can be performed if no further modifications of the polymer prior to hydrolysis into PG are needed.

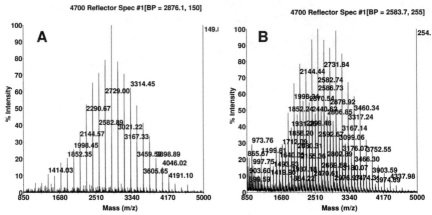

Figure 14   MALDI-TOF-MS spectra of polymer **15** obtained in the conditions corresponding to Table 7, Entry 3 before (A) and after (B) chromatographic purification on silica gel.

Purification of PBnGE was successfully performed chromatographically on silica gel as well.

4.1.5 Synthesis of Block Copolymers

The synthesized functional epoxides (Chapter 4.1.4.1), which are stable in anionic polymerization, can be polymerized into corresponding block copolymers with EO when PEG is used as macroinitiator, or when stepwise polymerization of EO followed by polymerization of functional epoxides is performed. Scheme 19 represents the range of obtained copolymers. The following epoxides were used for copolymer synthesis: AGE, EEGE and GADEA. The results are summarized in Table 8.

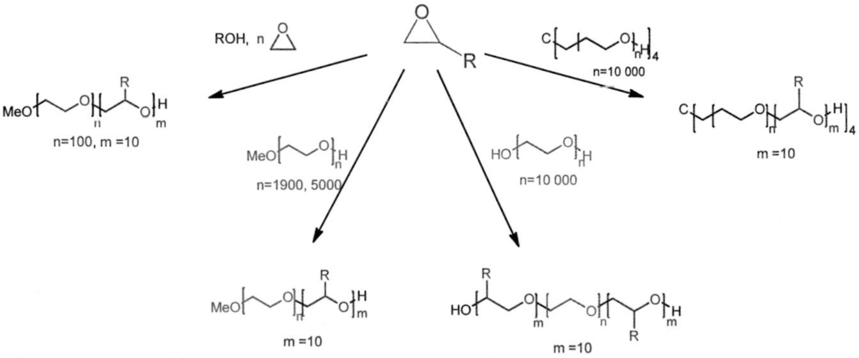

Scheme 19   Synthesis of block copolymers of functional epoxides (AGE, EEGE and GADEA) with EO and mPEGs/PEGs correspondingly.

Table 8   Synthesis and characterization of block copolymers of EO and functional epoxides.

entry	P^a	M^b	macroinitiator	I:M feed ratio^c	t (°C)	time^d (days)	conv.^{g} (%)	$\overline{DP_n}^h$	M_n (g/mol)	M_w (g/mol)	PDI^i
							^{1}H NMR^e		GPC^f		
1	17		mPEG 1900	1:10	rt	4	97	10	3800	3900	1.03
2	18	AGE	mPEG 5000	1:10	rt	4	95	9.5	7500	7700	1.03
3	19		PEG 10000	1:100	30	2	78	–	13700	14600	1.05
4	20		PEG 10000	1:200	30	7	97	–	21000	22000	1.05
5	21		mPEG 1900	1:10	rt	3	91	8.5	4200	4300	1.03
6	22	EEGE	mPEG 5000	1:10	rt	5	97	10	8100	8300	1.03
7	23		PEG 10000	1:10	rt	3	97	–	12300	12900	1.05
8	24		4-arms PEG 10000	1:40	rt	5	85	–	12200	12900	1.06
9	25	GADEA	2-methoxyethanol + 100 EO^j	1:10	rt	5	58	5.5	6000	6400	1.07
10	26		mPEG 1900	1:10	30	3	80	8	3600	3800	1.04

^{a}Polymer. ^{b}Monomer. ^{c}The feed ratio of macroinitiator to monomer. ^{d}Polymerization time. ^{e}Calculated based on ^{1}H NMR spectra. ^{f}Determined by GPC in THF relative to PS standards. ^{g}Conversion of monomer. ^{h}Degree of polymerization. ^{i}Polydispersity index = M_w/M_n. ^{j}Block copolymer was synthesized by stepwise polymerization of EO (2 day) and then GADEA (3 days) in a living manner. Initiator: 2-methoxyethanol/PN. The feed ratio of initiator to EO was 1:100.

Macroinitiators were prepared from mPEGs and PEGs of different molecular weights and PN similarly to the procedure described for the synthesis of homopolymers (Chapters 4.1.4.2.2 and 4.1.4.2.3). Diblock copolymers were obtained when mPEG was used as macroinitiator, while initiation with PEG resulted in triblock copolymers. Star-like block copolymer was synthesized by initiation with 4-arms PEG (Table 8, Entry 8). The fact that EO polymerization proceeds as living polymerization can be used for stepwise synthesis of block copolymers with the desired length of each block. Thus block copolymer mPEG-*b*-GADEA (**25**) was prepared by first polymerization of EO followed by addition of GADEA to the living PEG anions (Table 8, Entry 9). Because of the high viscosity of PEG anion solutions in THF the amount of solvent used for polymerization had to be increased in comparison with initiation by small molecular weight alkoxides. This resulted in lower conversions unless the polymerization has been performed for a longer time. In all cases obtained copolymers had low PDIs (Table 8).

The M_n values of obtained copolymers determined by GPC do not exactly correspond to the calculated ones (Table 8), which can be explained by the use of PS calibration for the measurement. Since the chemical structure of copolymers is different from the one of PS, it can be expected that the corresponding hydrodynamic volumes will also be different. Moreover, hydrodynamic volume of a macromolecule is also dependent on the architecture and size of the blocks. Therefore, it is impossible to find an appropriate calibration system and only general comparison of the obtained molecular weights as well as estimation of PDIs can be done. Comparative chromatograms of PEG 10000 and copolymer **20** (Figure 15A) show a decrease of retention time of PAGE-*b*-PEG-*b*-PAGE in comparison to initial PEG, which corresponds to an increase of molecular weight. The GPC measurements in water solutions with PEG standards could not be performed since block copolymers of EO and functional epoxides form micelles in water.[143]

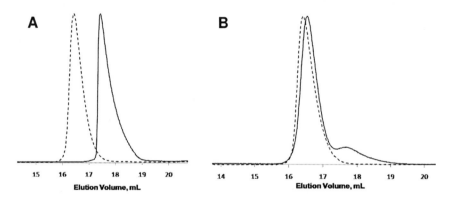

Figure 15   Gel permeation chromatograms of PEG 10000 (solid line) and polymer **20** (dashed line) obtained in the conditions corresponding to Table 8, Entry 4 (A), and of this polymer before (solid line) and after (dashed line) chromatographic purification on silica gel (B).

It should be noted that in all copolymerization reactions, small quantities of homopolymers were detected. This is due to the high hydrophilicity of PEGs and difficulties to completely remove water traces from the initiating system. Thus hydroxyl anion can also initiate polymerization leading to homopolymer formation. However, thanks to the significant difference in R$_f$ values of homomultifunctional PEGs and their block copolymer with PEGs, the latter can be easily purified chromatographically on silica gel (Figures 15B and 16).

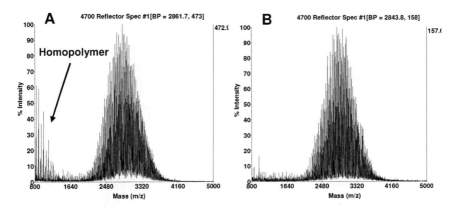

Figure 16   MALDI-TOF-MS mass spectra of block copolymer **17** obtained in the conditions corresponding to Table 8, Entry 1 before (A) and after (B) chromatographic purification on silica gel.

4.1.6   Post-Polymerization Cleavage of Protecting Groups in Multifunctional PEGs

PEEGE and GADEA are precursors of PG and multifunctional aldehyde PEG correspondingly and represent the case when protecting group had to be used in order to perform anionic polymerization of epoxide with the desired functionality. In both cases the protection of functional group (hydroxyl or aldehyde group) was done via acetal formation. Acetal protection is stable under basic conditions and can be cleaved upon treatment with acid.[73,77]

In case of PEEGE full deprotection took place after stirring for 10 min with hydrochloric acid. This was confirmed by ^{1}H NMR and MALDI-TOF measurements (Figure 17).

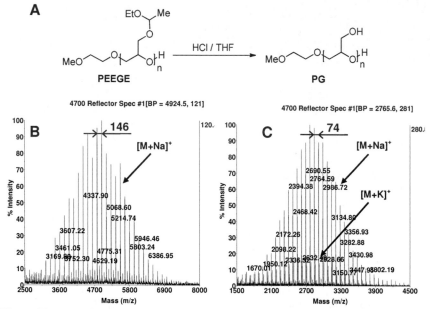

Figure 17   PEEGE deprotection into corresponding PG upon acid treatment (A) and MALDI-TOF-MS spectra of PEEGE (**15**) obtained in the conditions corresponding to Table 7, Entry 4 before (B) and after (C) deprotection.

Unfortunately, acid treatment of PGADEA in the same conditions as for PEEGE led to just partial cleavage of acetal protecting groups. Increase of the reaction time as well as use of trifluoroacetic acid instead of hydrochloric acid resulted in a complicated mixture of products. This could be explained by the numerous Aldol reactions taking place under acidic condition, being the reason for the instability of reactive aldehyde moiety. *In situ* deprotection – reaction with hydrazide led to no further improvement. This showed that multifunctional aldehyde PEG can not be obtained when acid assisted cleavage of protecting groups is necessary. Instead, milder conditions of deprotection should be applied.

4.2 Biomedical applications of Functional Epoxides and Multifunctional PEGs

4.2.1 Surface Modification with Functional Epoxides

4.2.1.1 Overview

Epoxides as Functional Linkers

Functional surfaces find various applications[170,171] including biosensors, tissue engineering, microfluidics and cell culture. For example, patterning of surfaces in a specific manner is highly desirable, as it is known that cells adhere better on rough surfaces.[172] Moreover, it was shown that cells behave differently depending on the size and type of surface pattern.[5,173] There are different techniques to introduce surface patterns including microcontact printing,[174,175] photolithography[176,177,178] and dip-pen nanolithography.[179,180,181] In this work, the microcontact printing technique was used for its simplicity and because the patterned stamps were readily available.

Functional epoxides can be used as linkers for surface modification, and to introduce new functionalities for further conjugation of biomolecules (Scheme 20).

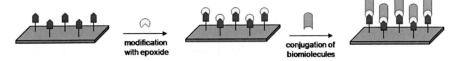

Scheme 20   Schematic representation of surface modification via epoxide chemistry.

Epoxides are reactive toward different types of nucleophiles, resulting in a ring opening (Scheme 21). If the surface bears any kind of nucleophilic groups such as hydroxyl or amino, then microcontact printing of a functional epoxide should give a modified surface with new chemistry. This could be a new versatile method for an easy and fast modification of surfaces with variety of chemical groups, since epoxides with different functionalities can be prepared (chapter 4.1.4.1).

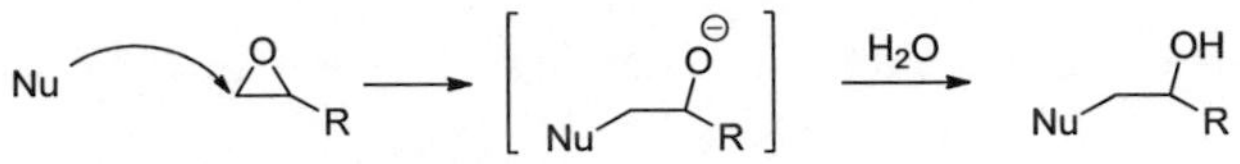

Scheme 21   Opening of the epoxide ring by nucleophiles.

Copper-Free Click Chemistry

Among other functional epoxides the one bearing an alkyne moiety is of particular importance since alkynes readily react with azides in a quantitative and highly specific manner. Because of these characteristics, Huisgen azide-alkyne cycloaddition has been classified as a click reaction and successfully employed in the fabrication of different biomedical materials and devices.[122,182] Mild conditions, absence of accompanying side processes and the possibility of performing reactions in water solutions distinguish azide-alkyne cycloaddition from other reactions when applied to biomolecules. However the reaction originally utilized copper salts, which were toxic for cells, requiring thorough washing procedures in order to remove any traces of copper after the reaction. To overcome this problem, a copper-free variant of the azide-alkyne cycloaddition was developed employing ring-strained alkynes.[183] However, their synthesis is typically a time-consuming multi-step procedure with quite low yields. On the other hand, several studies have been reported, utilizing alkynes with an electron withdrawing group close to the triple bond to perform copper-free Huisgen reactions.[184,185] Although the efficiency of these reactions was lower in comparison with sterically strained alkynes, the microcontact printing technique may facilitate the reaction, inducing the effect of electron withdrawing group.

4.2.1.2   Surface Modification with Reactive Epoxide for Copper-Free Click Chemistry

For the following surface modification, epoxide **27** with an electron deficient triple bond has been synthesized via esterification of glycidol (Scheme 22).

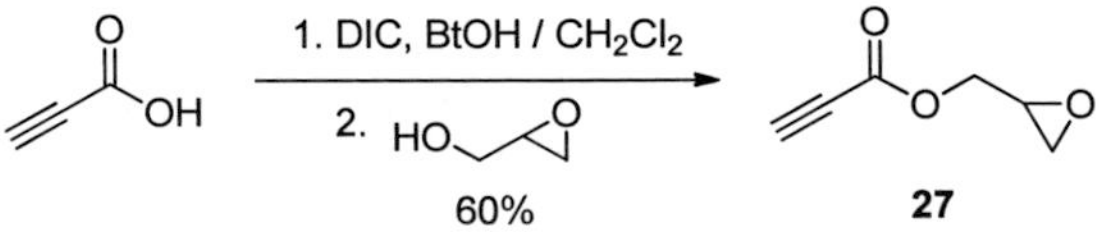

Scheme 22   Synthesis of an epoxide with an electron deficient alkyne group.

The silica or gold wafers used for this study were immobilized with amino functionalities via chemical vapor deposition (CVD) of amino paracyclophanes.[186] Further modification with epoxide **27** was performed in two ways (Figure 18A):

-   microcontact printing of epoxide **27** on the amino functionalized surfaces with flat stamp,
-   incubation of the amino functionalized wafers in the solution of the epoxide **27**.

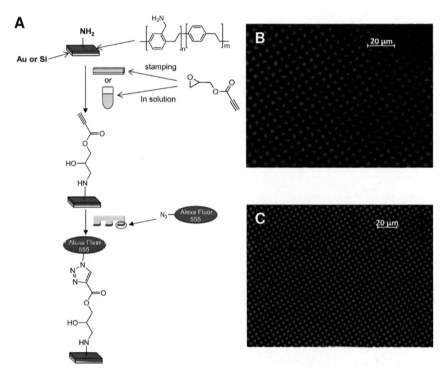

Figure 18   Schematic representation of the modification of amino functionalized surfaces with epoxide **27** bearing reactive alkyne group followed by immobilization of Alexa Fluor 555 azide (A) and fluorescent microscopy images of thus functionalized surfaces for the case of stamping epoxide **27** with a flat stamp (B) or incubation with epoxide **27** solution (C).

Surfaces modified with epoxide **27** by stamping with a flat stamp or in solution were analyzed by X-ray photoelectron spectroscopy (XPS) (Table 9). The values for C, N, and O content on the analyzed substrates are closer to the calculated values in the case of modification in solution, meaning that this method is more effective in comparison with the stamping technique. However, both methods give satisfactory results and can be further applied.

Table 9   XPS analysis of the surfaces after modification with epoxide **27** either by printing with a flat stamp or by incubation in the epoxide solution.

| | XPS analysis | | |
atom	calculated (%)	flat stamp[a] (%)	solution[b] (%)
C	85.2	90.3	87.0
N	3.7	4.4	3.8
O	11.1	5.3	9.1

[a]Surfaces modification was done by microcontact printing of epoxide **27** with a flat stamp. [b]Surface modification was done by incubation of wafers in the epoxide **27** solution.

Next, azide conjugated red fluorescent dye Alexa Fluor 555 was printed on the surface with square (3×3 μm) featured stamp. After a washing procedure the samples were imaged by fluorescent microscopy (Figures 18B and C). The dye binds to the surface due to reaction of its azide group with activated alkyne groups on the surface.

To expand these experiments for the binding of biomolecules, microcontact printing of azide conjugated biotin with square (3×3 μm) featured stamp and line (width 40 μm) featured stamp on the epoxide **27** modified surfaces was performed (Figure 19). Since biotin has a very high affinity for steptavidin, the latter, when labeled with a dye, can be used for visualization of biotin. After incubation of biotine modified surfaces with rhodamine-labeled streptavidin (RB-streptavidin) it could be observed in fluorescent microscopy images that fluorescence appears specifically in the squares or lines – the areas where biotin was previously stamped.

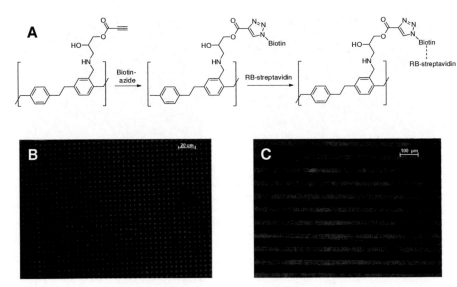

Figure 19   Schematic representation of biotin-azide immobilization on the surfaces functionalized with epoxide **27** followed by RB-steptavidin conjugation (A) and fluorescent microscopy images of thus functionalized surfaces (biotin-azide stamped with square (3×3 μm) featured stamp (B), with line (width 40 μm) featured stamp (C)).

Unfortunately, the attempts to change the sequence of stamping and first immobilize epoxide **27** with a featured stamp and then conjugate Alexa Fluor 555 azide in solution or by printing with flat stamp did not give satisfying results (Figure 20). This is likely explained by high volatility of epoxide **27**, which has a very low molecular weight. The features of the stamp were very small meaning that the amount of the epoxide solution introduced on each square was extremely small and could evaporate in a very short time. The introduction of epoxide **27** was done in solution and evaporation of the solvent (to ensure that only features of the stamp would come in contact with the modified surface) required some time. This time was probably enough for the epoxide to evaporate from the surface.

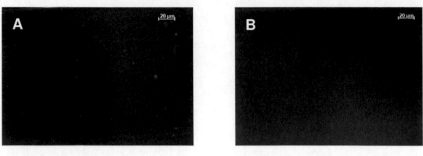

Figure 20   Fluorescent microscopy images of the reversed modified surfaces: first immobilization of epoxide **27** with a square (3×3 μm) featured stamp, then conjugation of Alexa Fluor 555 azide in solution (A) or by printing with flat stamp (B).

Finally a control experiment was undertaken by printing Alexa Fluor 555 azide directly on the amino functional surface (without preliminary modification with epoxide). Unfortunately, this experiment resulted in unspecific reaction of the dye with amino groups of the surface (Figure 21).

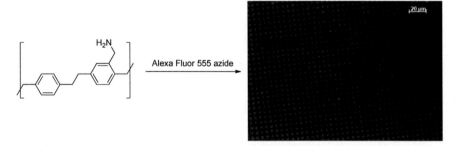

Figure 21   Fluorescent microscopy image of the Alexa Fluor 555 azide direct printing on the amino functional surface with a square (3×3 μm) featured stamp. The appearance of fluoresce in the stamped areas indicates unspecific reaction.

To avoid unspecific reactions, stepwise modification of the surface can be completed first introducing the reactive species and next passivating the rest of the surface with non-reactive molecules.

4.2.2 Reactive Hydrogels Based on Multifunctional Photosensitive Protected Aldehyde PEG

The materials in this chapter have been adapted with minor modifications from the paper currently in preparation: E. Sokolovskaya, S. Bräse, J. Lahann. *Synthesis and On-Demand Gelation of Multifunctional Poly(ethylene glycol) Derivatives.*

4.2.2.1 Overview

Hydrogels

Hydrogels are highly hydrophilic polymer networks, which are not water soluble, but are able to absorb large amounts of water.[187] Due to their hydrophilicity and similarity of their mechanical properties to those of biological soft tissues, hydrogels have found diverse applications in biomedicine[53,188,189] for drug delivery, tissue engineering, biosensors and diagnostic imaging. Some examples of PEG-based hydrogels applications were described in the introduction section (Chapter 2.1.3).

Specific characteristics of hydrogels determine their properties and future application. The mechanical strength and stability of a hydrogel is primarily defined by degree and type of crosslinking. Covalent bonding results in the formation of stable hydrogels, while crosslinking based on physical interactions (ionic and hydrogen bonding or hydrophobic interactions) gives less stable hydrogels with weaker mechanical properties, though allowing reversibility in the hydrogel's gelling properties.[188]

An important aspect is the biodegradability of a hydrogel. The strategies to create degradable hydrogels can rely either on the nature of polymer (degradable synthetic or biopolymers) or on the type of crosslinking (labile bonds). If degradable bonds (within a polymer or crosslinking bonds) are designed in a manner in which they can be cleaved upon action of some stimuli, then new responsive hydrogels can be prepared.

Specific applications of hydrogels may lead to additional requirements for the gel formation. Thus, hydrogels designed as carriers of bioactive molecules (proteins, genes, etc.) or scaffolds for cell encapsulation require mild and close to physiological conditions of gel formation.[54] This can be achieved, for example, by means of photoinduced gel formation. Furthermore, recently there have been increased affords in generating systems suitable for *in situ* gelation. With the use of these systems, loading of biomolecules can be achieved simply by their dissolution in the gel forming polymer solution, followed by gelation at the injection site.[190,191] Thus introduction of a hydrogel into human body can be performed with minimum invasion. *In situ* gelation by photopolymerization results in the formation of stable bonds; therefore if the polymer is not degradable itself, then the units providing degradation should be introduced additionally. Chemical crosslinking commonly relies on click chemistries, such that gel formation takes place simply upon mixing of two gel components. Thus aldehyde/hydrazide chemistry was realized by reaction of oxidized dextrane and adipic acid dihydrazide, forming acid sensitive hydrazone linkage readily degradable at low pH.[192] This is an excellent example of the use of click chemistry resulting in pH responsive degradable hydrogel. PEG based multifunctional aldehyde polymers can be used instead of dextrane; thus the oxidation step can be avoided. Furthermore, protection of the aldehyde with a photolabile group would allow for control of gel formation by applying UV light. Such polymers can be obtained by polymerization of the corresponding epoxide bearing photosensitive protected aldehyde (PPA) group.

Photolabile Protecting Groups for Carbonyl Compounds

Photolabile protecting groups were developed for a wide range of functional groups such as hydroxyl, carboxyl and amino groups.[193,194,195] However there are not many examples of existing photolabile protecting groups for aldehydes. Recently Wang et al. reported on the synthesis of a new photolabile carbonyl protecting group based on a salicyl alcohol derivative (Scheme 23).[196]

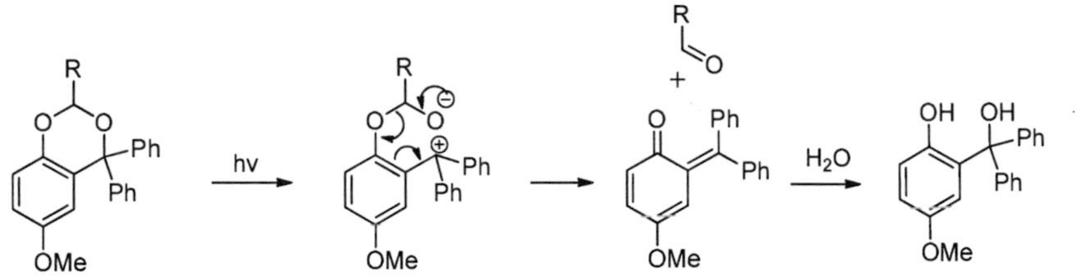

Scheme 23   Mechanism of photoinduced cleavage of aldehyde protecting group.[196]

As deprotection takes place under neutral conditions upon irradiation, it can be a suitable alternative to the previously used diethoxy acetal protection in the synthesis of PGADEA, which is removed under acidic condition thus causing various side condensation reactions (Chapter 4.1.6).

On the other hand, the possibility to trigger the cleavage of aldehyde protecting groups by exposition to UV light allows for the control over the following reactions of thus released aldehyde groups and, as it has been suggested above, can be utilized for controlled formation of a hydrogel.

4.2.2.2 Synthesis of an Epoxide with PPA Group

The synthesis of an epoxide with PPA group was performed as shown in Scheme 24. Diol **30** was synthesized similar to the procedure previously reported by Wang et al.,[196] except that for the Grignard reaction acid **28** was preliminary converted into corresponding ester **29**, because of the very low yield obtained when acid **28** was directly reacted with phenylmagnesium bromide.

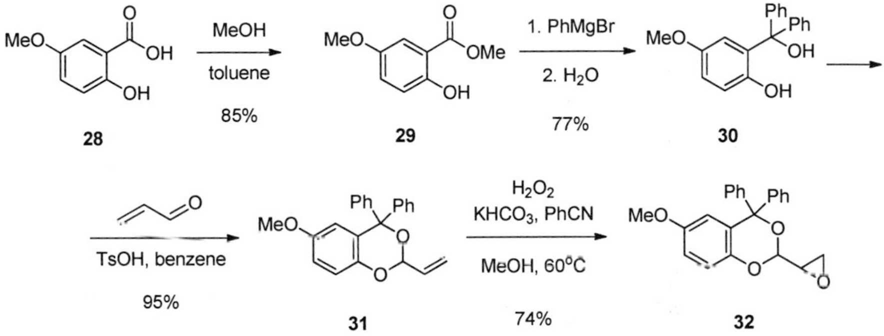

Scheme 24   Synthesis of an epoxide with PPA group.

Reaction of diol **30** with acrolein under acidic catalysis resulted in the formation of photosensitive acetal **31**. Further epoxidation of the double bond yielded the target epoxide **32**. In each step a product was purified chromatographically on silica gel. It should be noted that epoxide **32** has two asymmetric carbons and was obtained as a mixture of two diastereomers in the ratio 1:1.35. It is possible to separate the diastereomers by column chromatography, however, their R_f values are very close, which complicates the separation and results in low yields of pure diastereomers. Therefore for all further experiments epoxide **32** was used as a mixture of diastereomers.

4.2.2.3 Synthesis of Multifunctional PPAPEG

Polymerization of epoxide **32** was performed as described before for the polymerization of functional epoxides in Chapter 4.1.4.2.3 (Scheme 25).

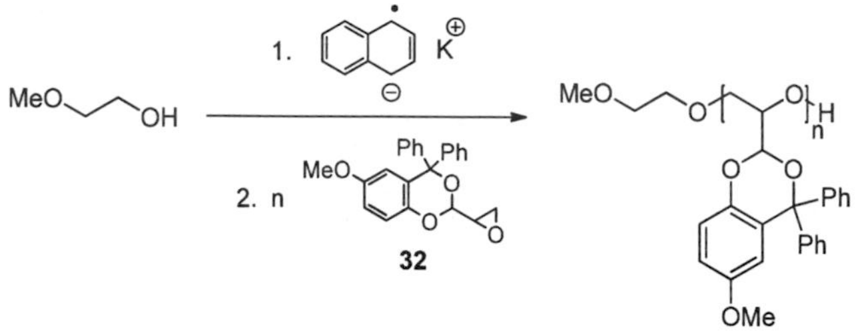

Scheme 25   Synthesis of multifunctional PPAPEG by anionic ring opening polymerization of the corresponding epoxide.

Since epoxide **32** is a solid compound, polymerization was done in THF solution. When the polymerization was performed at room temperature no polymer was detected in the reaction mixture (Table 10, Entry 1). This can be explained by the presence of a very bulky substituent hindering access to the epoxide ring. Increasing the polymerization temperature and reaction time made the synthesis of polymer **33** possible (Table 10, Entry 2).

Table 10   Conditions for anionic polymerization of epoxide **32** and characterization of the obtained polymers.

entry	I:M feed ratio[a]	t (°C)	Polymerization time (days)	Monomer conv.[b] (%)	GPC[c]		
					M_n (g/mol)	M_w (g/mol)	PDI[d]
1	1:8	rt	3	0	–	–	–
2	1:6	40	5	86	900	1300	1.48
3	1:44	40	6	45	2600	3700	1.43
4	1:44	50	6	75	2800	4500	1.61
5	1:100	50	8	49	4300	5800	1.36

[a]The feed ratio of initiator to monomer. [b]Calculated based on ^{1}H NMR spectra. [c]Determined by GPC in THF relative to PS standards. [d]Polydispersity index = M_w/M_n.

Unfortunately, the conversions were quite low with increased number of monomer equivalents even when the polymerization temperature was raised to 50°C (Table 10, Entries 3–5). Molecular weights of the obtained polymers determined by GPC were considerably lower than those calculated based on the monomer conversions. However, this could be partially because of the use of PS standards, which did not adequately reflect the hydrodynamic volume of polymer **33** in THF solution. On the other hand, long reaction time and temperatures above 30°C could facilitate the transfer reactions. This could also be the reason for higher PDIs as compared to those of previously obtained functional PEGs (Chapters 4.1.4.2.2 and 4.1.4.2.3). However, high PDIs values could also be due to the high molecular weight of the monomer and as a result a broader distribution of molecular weights of the polymeric chains differed by just one or two monomer units. This is particularly true in this case as the obtained polymers in general had low molecular weights.

As was previously shown in the case of epoxide **32**, photolabile acetal protection was stable during chromatographic purification on silica gel (Chapter 4.2.2.2). Thus chromatographic purification of PPAPEG should also be possible. Indeed, as seen in the GPC profiles of the crude and pure polymers (Figure 22), the monomer peak completely disappears after purification and at the same time the molecular weight of the polymer is not changed. This confirms the stability of the protection group on silica gel.

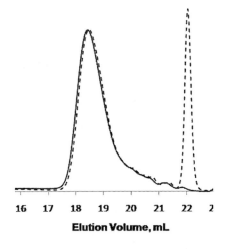

Elution Volume, mL

Figure 22   Gel permeation chromatograms of polymer **33** obtained in the conditions corresponding to Table 10, Entry 4 before (dashed line) and after (solid line) chromatographic purification on silica gel.

4.2.2.4 Photoinduced Cleavage of Aldehyde Protecting Group

According to the UV spectrum of PPAPEG, its maximum absorption is 297 nm while at 365 nm the polymer does not absorb at all (Figure 23). Since absorption at 312 nm is still sufficient (40%), and UV lamps with maximum intensity at 312 nm are commercially available, all further experiments involving removal of photolabile acetal groups were performed at this wavelength.

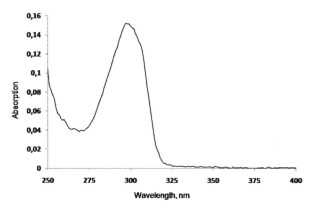

Figure 23   UV spectrum of PPAPEG.

To demonstrate the fact that cleavage of acetal protecting groups resulting in the release of multifunctional aldehyde PEG is taking place upon irradiation of PPAPEG, the polymer was spin coated on gold wafers and irradiated for different periods of time. The corresponding IR spectra are presented in the Figure 24. Appearance of a strong signal at 1730 cm^{-1} confirms presence of free aldehyde moieties. At the same time the intensity of the band at 1045 cm^{-1} corresponding to C–O–C stretching decreases which is in accordance with cleavage of acetal bonds. Finally, a pair of broad signals in the area 3100–3500 cm^{-1} are due to the hydroxyl groups of the released diol **30** (Scheme 23).

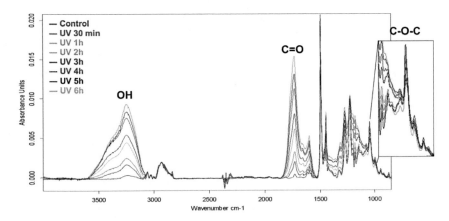

Figure 24   IR spectra of PPAPEG before (control) and after different times of irradiation with 312 nm light. PPAPEG was spin coated on the gold wafers.

Next, the reactivity of PPAPEG toward hydrazides upon irradiation was investigated. To simplify the system and get better peak assignment in ^{1}H NMR spectra, the low molecular weight analogue of PPAPEG – alcohol **34** – was prepared. The synthesis was performed similar to the polymerization of epoxide **32** (Scheme 25), but with excess of initiator. Like epoxide **32**, compound **34** has two asymmetric centers and was obtained as a mixture of two diastereomers in ratio 1:2.3. However, unlike epoxide **32** the R_f values of diastereomers **34** were sufficiently different to allow for successful separation by column chromatography. To avoid overlay of signals in NMR spectra further study was performed with the use of just one pure diastereomer.

Solution of **34** and acetohydrazide in CD$_3$CN:D$_2$O (9:1) was prepared and exposed to 312 nm light. This should lead to the cleavage of acetal protecting groups and formation of aldehyde **35**. The latter should quickly react with the present in the reaction mixture acetohydrazide to form isomeric hydrazones **36** (Scheme 26).

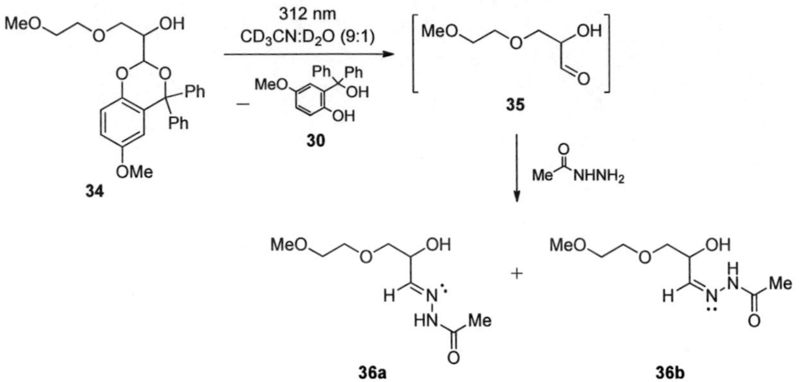

Scheme 26   Photoinduced deprotection of alcohol **34** upon irradiation with 312 nm light followed by *in situ* reaction with acetohydrazide.

Indeed after two hours irradiation of the prepared mixture with 312 nm light, complete disappearance of acetal was detected in ^{1}H NMR spectrum (Figure 25B). Instead there were two new multiplets at 4.33 and 4.27 ppm in ratio 1.7:1 correspondingly. These two multiplets should be assigned to the protons **a** of the isomeric hydrazones as shown in Figure 25B.

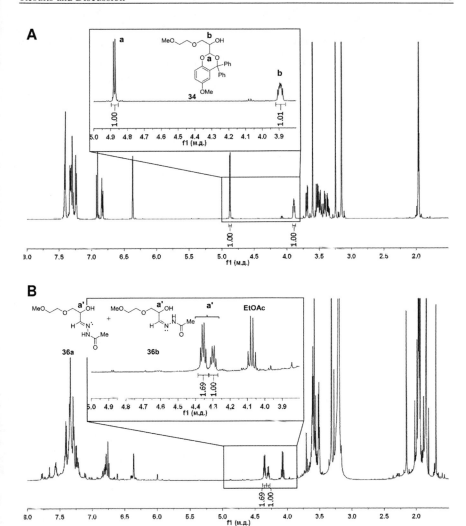

Figure 25 ^{1}H NMR spectra of **34** (A) and isomeric hydrazones **36** obtained by photodeprotection of **34** followed by *in situ* reaction with acetohydrazide (B). Solvent CD$_3$CN:D$_2$O (9:1).

4.2.2.5   Synthesis of Water Soluble PPAPEG-*b*-PEG-*b*-PPAPEG

Synthesis of hydrogels requires water solubility of the gel forming polymers. However, homomultifunctional PPAPEG is highly hydrophobic and reactions of this polymer in water solutions are not possible. Hydrophilic properties of a polymer can be varied when copolymers with different properties of monomer units are synthesized. PEG is often used to increase the solubility of small molecular weight compounds as well as to obtain water soluble polymers (Chapter 1). To get a water soluble analogue of PPAPEG a series of its block copolymers with PEG (**37**) has been prepared (Table 11) and their solubility in water has been tested.

Table 11   Synthesis and characterization of PPAPEG-*b*-PEG-*b*-PPAPEG block copolymers.

entry	polymer	PEG[a]	I:M feed ratio[b]	polymer.[c] time (days)	monomer conv.[d] (%)	M_n (g/mol)	M_w (g/mol)	DPI[f]
						GPC[e]		
1	**a**	20 000	1:200	6	65	18800	23100	1.23
2	**b**	20 000	1:30	6	51	13300	15400	1.16
3	**c**	20 000	1:3	3	99	14700	17100	1.16
4	**d**	3 000	1:2.4	5	99	4700	4 900	1.04

[a]M_n of PEG used as macroinitiator. [b]The feed ratio of macroinitiator to monomer. [c]Polymerization time. [d]Determined by ^{1}H NMR spectroscopy. [e]Determined by GPC in THF relative to PS standards. [f]Polydispersity index = M_w/M_n.

High molecular weight copolymer with long PPAPEG blocks (~65 units each block considering conversion) was completely water insoluble (Table 11, Entry 1). Decreasing the number of PPAEG units per block to ~ 7.5 units (considering conversion) still didn't yield a water soluble polymer (Table 11, entry 2). The latter more likely tends to form micelles in water solutions. Finally, when just one or two PPAEG units were attached to the PEG termini the copolymer was completely soluble in water (Table 11, Entry 3). Still copolymer **37c** is not an optimal candidate for hydrogels formation, since crosslinking would be possible just through the termini of the polymer, while the connecting PEG block would be too long. Thus crosslinking density would be low and hydrogel would have poor mechanical properties. To overcome this drawback, PEG 3000 was used as macroinitiator for the synthesis of PPAPEG-*b*-PEG-*b*-PPAPEG (Table 11, Entry 4, Figure 26A). After chromatographic purification on silica gel the obtained polymer **37d**, although giving cloudy solution, was soluble enough to perform hydrogel synthesis.

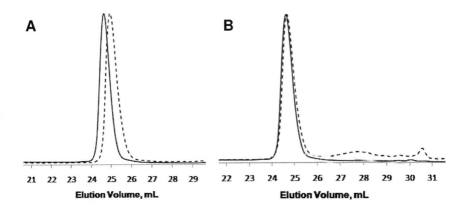

Figure 26   Comparative gel permeation chromatograms of PEG 3000 used as macroinitiator (dashed line) and PPAPEG-*b*-PEG-*b*-PPAPEG copolymer **37d** obtained in the conditions corresponding to Table 11, Entry 4 (solid line) (A) and this polymer before (dashed line) and after (solid line) chromatographic purification on silica gel (B).

Photoinduced deprotection of acetal groups in PPAPEG-*b*-PEG-*b*-PPAPEG followed by *in situ* reaction with acetohydrazide leads to the formation of the corresponding hydrazones (Scheme 27), which was confirmed by the appearance of corresponding signals (a') in ^{1}H NMR spectra (Figure 27).

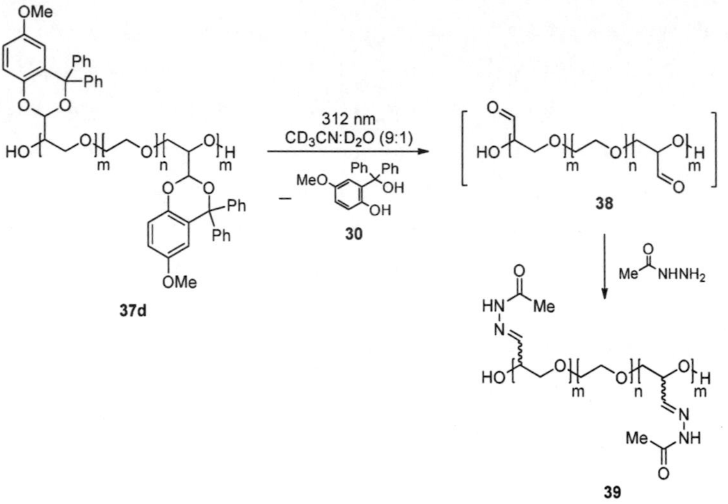

Scheme 27   Photodeprotection of PPAPEG-*b*-PEG-*b*-PPAPEG upon irradiation with 312 nm light followed by *in situ* reaction with acetohydrazide.

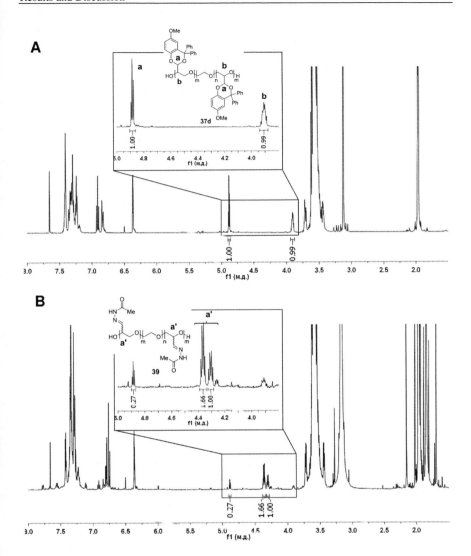

Figure 27   ¹H NMR spectra of PPAPEG-*b*-PEG-*b*-PPAPEG (A) and hydrazones obtained by photodeprotection of aldehyde groups in PPAPEG-*b*-PEG-*b*-PPAPEG followed by their *in situ* reaction with acetohydrazide (B). Solvent CD₃CN:D₂O (9:1).

4.2.2.6 Synthesis of Multifunctional Hydrazide PEG

In order to perform hydrogels synthesis based on PPAPEG-*b*-PEG-*b*-PPAPEG via aldehyde/hydrazide reaction, a second component – multifunctional hydrazide polymer – is necessary.

Synthesis of multifunctional hydrazide PEG (PHZ) via direct polymerization of corresponding epoxide is impossible. Therefore the post-polymerization approach should be applied.[121] The synthesis of linear PHZ (**41**) can be performed via a two step modification of PAGE (**12b**) (Scheme 28), similar to the synthesis of diblock copolymer mPEG-*b*-PHZ reported previously.[125]

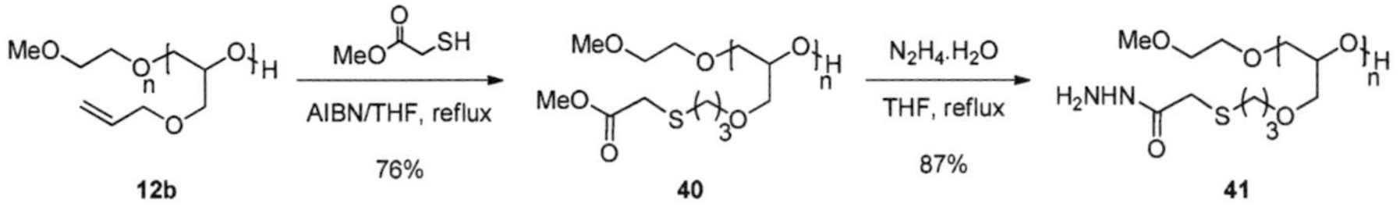

Scheme 28   Synthesis of homomultifunctional PHZ.

Modification of multifunctional polymers is preferably performed via click reactions.[122] These reactions in contrast to classic organic reactions proceed with quantitative conversions and without side reactions, which is crucial for multifunctional polymer modification, because the separation of unmodified starting moieties from a product macromolecule is not possible by any purification methods. Thiol-ene reaction is commonly classified as a click reaction,[66] though when applied for modification of alkene bearing polymers, excess thiol is needed to prevent radical initiated crosslinking of double bonds. PAGE (**12b**) was converted into the corresponding ester **40** via thiol-ene reaction with methyl mercaptoacetate. The excess of the thiol used in the reaction was successfully removed chromatographically, similar to the purification of other multifunctional PEGs (Chapter 4.1.4.2).

Purified polymer **40** was further reacted with an excess of hydrazine, yielding the desired PHZ (**41**). Full conversion of ester groups into hydrazide groups was confirmed by the disappearance of the singlet of methoxy groups in ^{1}H NMR spectrum of polymer **41** (Figure 28). Purification of PHZ from the excess of hydrazine hydrate was performed by dialysis. Comparison of MALDI-TOF-MS spectra of polymers **40** and **41** revealed no change in the molecular weight (Figure 29).

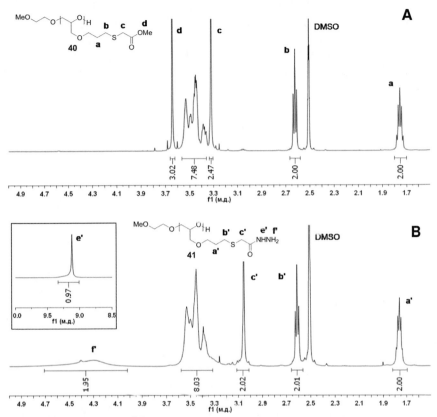

Figure 28   Comparative ¹H NMR spectra of polymers **40** (A) and **41** (B). Solvent DMSO-
d_6.

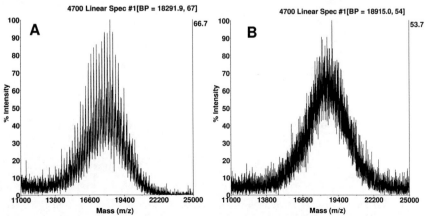

Figure 29   Comparative MALDI-TOF-MS spectra of polymers **40** (A) and **41** (B).

4.2.2.7   Hydrogels Fabrication

PPAPEG-*b*-PEG-*b*-PPAPEG and PHZ, synthesized as described above, were used for the photoinduced fabrication of a hydrogel. The process is schematically shown in Scheme 29.

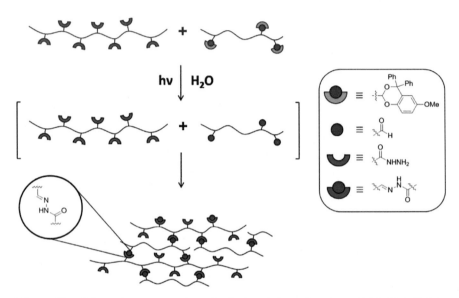

Scheme 29   Schematic representation of a hydrogel formation from PPAPEG-*b*-PEG-*b*-PPAPEG and PHZ in water upon irradiation with 312 nm light. Exposition of PPAPEG-*b*-PEG-*b*-PPAPEG to UV light causes cleavage of photolabile acetal protection releasing free aldehyde groups, which directly react with the hydrazide groups of PHZ forming hydrazone crosslinking bonds.

Both PPAPEG-*b*-PEG-*b*-PPAPEG and PHZ were dissolved in water to yield white cloudy solution with total concentration of polymers 10 w/v% (Figure 30A). For the first 30 min the solution was stirred while exposed to UV light. After first 2 min of irradiation solution became absolutely clear and colorless. After 5 min the solution turned yellow and the color intensified further with longer time of irradiation. After 20 min a phase separation could be observed, and after 30 min a clear gel formation was detected. To avoid physical destruction of the hydrogel, stirring was stopped and solution was left under UV light for additional 1.5 h. Total time of irradiation was chosen to be 2 h based on the previous experiments with compound **34**, which showed complete disappearance of acetal signals after 2 h. The obtained hydrogel is shown in Figure 30B and C.

This method to fabricate hydrogels allows for controlled gel formation under mild conditions of UV irradiation compatible with the loading of biomolecules or cells encapsulation and provides degradable hydrogels due to the presence of acid-labile hydrazone bonds. Thus prepared hydrogels have a great potential for further applications in biomedicine including drug delivery with possibility of *in situ* gelation and controlled release of encapsulated material.

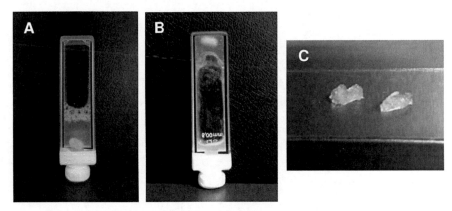

Figure 30   Formation of hydrogel from PPAPEG-*b*-PEG-*b*-PPAPEG and PHZ in water upon irradiation with 312 nm light. Images of the reaction mixture before (A) and after irradiation for 2 h (B, C).

4.2.3   Multifunctional Hydrazide PEG and Sugar Reactive Particles

The materials in this chapter have been adapted with minor modifications from the following paper: E. Sokolovskaya, J. Yoon, A. C. Misra, S. Bräse, J. Lahann, *Macromol. Rapid Commun.* **2013**, accepted. *Controlled Microstructuring of Janus Particles Based on a Multifunctional Poly(ethylene glycol).*

4.2.3.1   Overview

Applications of Nano and Micro Particles

The demand for nano and microparticles in biomedicine,[197] electronics,[198] optics[199] and catalysis[200] has grown dramatically in recent years due to the discovery of specific properties and functions imparted by their size. In biomedicine, great progress has been achieved with the use of nano/micro particles related to drug/gene delivery,[21] tissue engineering[201] and medical diagnostics.[202] Chemical composition as well as physical characteristics such as size, shape and mechanical properties of particles are important factors as they greatly influence particle interactions with an organism, circulation time in blood, cellular uptake and the ability to carry and release molecules.[21,203] In addition, fabrication of anisotropic particles offers interesting perspectives for advanced drug delivery allowing for delivery combined with imaging[49] or targeting,[48] or the combination of drugs with different release kinetics.[50]

Methods to Fabricate Nano and Micro Particles

Various techniques are known to create anisotropic particles with specific properties,[204] including matrix embedment, pickering emulsion polymerization, glancing angle deposition, microfluidics and electrohydrodynamic (EHD) co-jetting. Great success was achieved in introducing surface anisotropy by generation of so-called "patchy" particles.[205] However, the methods used to create particles with bulk anisotropy are basically limited to microfluidics and EHD co-jetting. The latter is a highly reproducible process, which allows for variation of several particle parameters, including size and shape, and introduction of anisotropy by fabrication of multicompartmental particles.[206]

In a typical setup of EHD co-jetting (Scheme 30), two jetting solutions are pumped through the side-by-side capillary system in a laminar flow. The applied electrical field causes distortion of a droplet at the tip of the nozzle into a Taylor cone and the formation of a very thin polymer thread. Ultrafast solvent evaporation results in precipitation and solidification of the particles before they reach the substrate surface. Particles or fibers can thus be fabricated depending on jetting conditions such as polymer concentration in the jetting solutions and flow rate.[207]

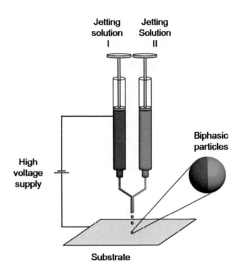

Scheme 30   Schematic representation of electrohydrodynamic co-jetting technique for the generation of anisotropic particles.

Polymers for Fabrication of Nano and Microparticles by EHD Co-jetting

Various polymers have been used thus far for EHD co-jetting. The choice of the polymer is usually dictated by the future application, which is why for biomedical applications the most vital criteria for polymer selection are biocompatibility and biodegradability. Poly(lactic-*co*-glycolic acid (PLGA) is most commonly used for this purpose as it satisfies the aforementioned criteria and is commercially available.[207] PEG, PAA and polyethylene imine (PEI) have also been used for jetting in water solutions,[208] however EHD co-jetting in water solutions requires further crosslinking of the generated particles in order to avoid their dissolution under physiological conditions. This can be done with the use of chemical crosslinkers and photo- or thermal crosslinking reactions. However, these reactions may augment the properties of the molecules loaded in the microparticles. Thus the jetting of water insoluble polymers in organic solutions is preferred. The drawback of the particles based on commercially available polymers is the lack of functional groups, which can be used for further particle modification and conjugation of targeting molecules. As a result, there is a need to search for new functional polymers.

Multifunctional PEGs offer a new synthetic platform for fabrication of functional anisotropic particles (Chapter 2.2.3). Among others, PEG with a multitude of hydrazide groups is particularly attractive as hydrazide groups are known to react specifically with aldehydes at ambient conditions, forming acid sensitive hydrazone linkages. This has potential applications for the delivery of aldehyde containing drugs[125] or sugar-conjugation for glucotargeting.[209] A block copolymer of mPEG and multifunctional hydrazide PEG (mPEG-*b*-PHZ) has been synthesized by Hruby et al. and used for the doxorubicin conjugation and fabrication of micelles carriers.[125] Later Zhou et al. modified this delivery system by synthesizing an analogous random copolymer P(EG-*co*-HZ), which after immobilization of doxorubicin resulted in a water soluble conjugate.[210] However, these reported polymers would not be suitable for jetting in organic solvents because of their hydrophilic properties.

In Chapter 4.2.2.6 the synthesis of homomultifunctional PHZ according to procedure by Hruby et al. was presented. This homomultifunctional polymer was also water soluble and did not satisfy the jetting conditions. Increasing the hydrophobicity of the polymer can be done via the introduction of hydrophobic moieties, such as alkane or phenyl groups, into the polymer chain. Such water insoluble multifunctional hydrazide PEG would be suitable for EHD co-jetting and could be used for fabrication of sugar-conjugated drug carriers.

4.2.3.2 Synthesis of Water Insoluble Multifunctional P(HZ-*co*-BnGE)

For the synthesis of hydrophobic PHZ analogue two possible approaches are applicable: 1) post-polymerization modification of homomultifunctional PAGE via two sequential thiol-ene reactions, introducing first ester groups for further conversion into hydrazide groups and then hydrophobic groups; 2) copolymerization of AGE with epoxide bearing hydrophobic group (alkane or phenyl) followed by subsequent modification into hydrazide polymer.

The first approach was realized by reaction of PAGE (**12b**) with methyl mercaptoacetate followed by *in situ* reaction of partially modified PAGE (**42**) with benzyl mercaptan (Scheme 31). Polyester **43** was obtained with the ester to benzyl groups ratio equal to 1:1.7 and after purification reacted with excess of hydrazine hydrate to yield polyhydrazide **44**.

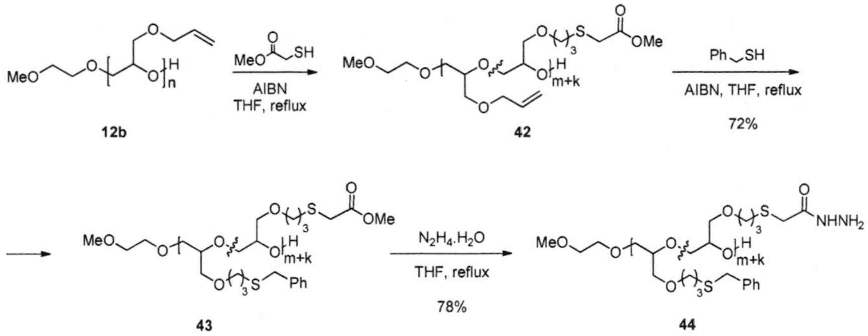

Scheme 31   Synthesis of water insoluble multifunctional hydrazide PEG via approach 1.

The drawback of this approach are the side reactions of allyl groups leading to crosslinking and consequently to a broad molecular weight distribution of the polymer **43** (PDI 1.26). Typically this problem is solved by the use of thiol excess. However the first step thiol-ene reaction has to be performed in the lack of thiol to avoid full consumption of allyl groups.

The second approach utilizing a copolymer of AGE and BnGE as a starting polymer does not have this drawback. For this reason, this method was further used for the preparation of hydrophobic PHZ analogue (Scheme 32).

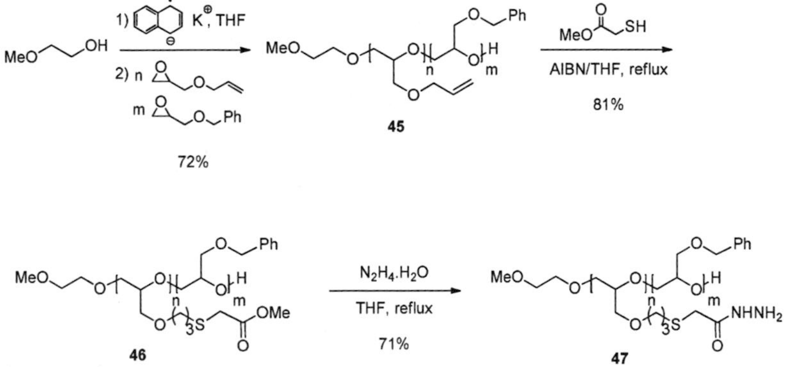

Scheme 32   Synthesis of water insoluble multifunctional hydrazide PEG via approach 2.

To ensure water insolubility of the final P(HZ-*co*-BnGE) (**47**) while maintaining a sufficient number of hydrazide groups per polymer chain, the ratio of comonomers AGE:BnGE was chosen to be 1:2. The copolymer P(AGE-*co*-BnGE) (**45**) was obtained with high monomer conversions (99% and 97% for AGE and BnGE correspondingly) and low polydispersity index (1.07). After chromatographic purification on silica gel polymer **45** was further converted into corresponding polyhydrazide **47** (Scheme 32) following the procedure for the synthesis of PHZ (Chapter 4.2.2.6). GPC chromatograms of polymers **45** and **46** are shown in Figure 31. Full conversion of ester groups into hydrazide groups was confirmed by the disappearance of the singlet of the methoxy groups in the ^{1}H NMR spectrum of the polymer **47** (Figure 32). Since polymer **47** was soluble in organic solvents, the isolation from the excess of hydrazine hydrate was performed by simply washing the polymer solution in chloroform with water. Comparison of MALDI-TOF-MS spectra of polymers **46** and **47** revealed no change in the molecular weight (Figure 33). Moreover, after modifications the ratio of comonomers in polymer **47** remained 1:2, as calculated from the ^{1}H NMR spectrum of polymer **47** (Figure 32B).

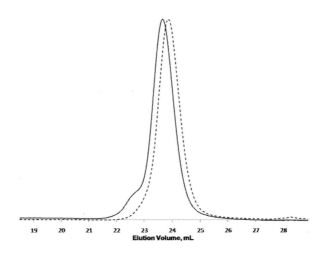

Figure 31   Comparative gel permeation chromatograms of polymers **45** (dashed line) and **46** (solid line).

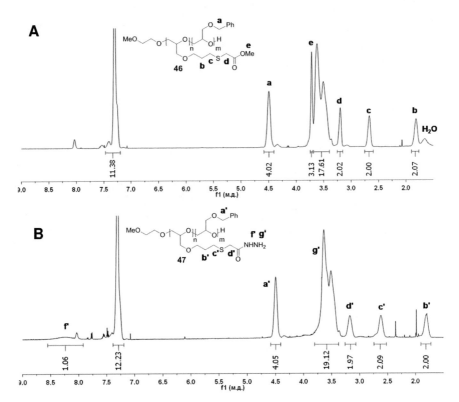

Figure 32   Comparative ^{1}H NMR spectra of polymers **46** (A) and **47** (B).

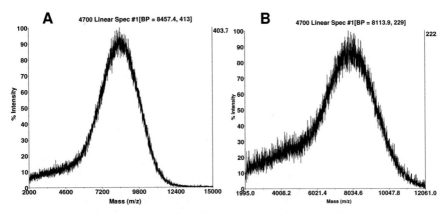

Figure 33   Comparative MALDI-TOF-MS spectra of polymers **46** (A) and **47** (B).

The obtained P(HZ-*co*-BnGE) (**47**) showed very low solubility in water, approximately 0.4 μg/mL, as determined visually. By comparison, the solubility of an analogous homomultifunctional PHZ was estimated to be equal to 0.5 mg/ml, which is more than three orders of magnitude higher than that of P(HZ-*co*-BnGE) (the representative solutions of P(HZ-*co*-BnGE) and PHZ are shown in Figure 34).

Figure 34   Comparison of the solubility in water of PHZ (0.5 mg/mL) (A) and P(HZ-*co*-BnGE) (5μg/mL) (B). Image of pure water is shown for comparison (C).

4.2.3.3   Fabrication and Modification of P(HZ-*co*-BnGE) Microparticles

Fabrication and modification of P(HZ-co-BnGE) microparticles was completed by Dr. Jaewon Yoon.

P(HZ-co-BnGE) Microspheres

To investigate potential uses of these multifunctional polymers, this work was expanded to the development of bicompartmental microspheres and microcylinders having polymer P(HZ-*co*-BnGE) selectively in one of the compartments. First, PLGA/P(HZ-*co*-BnGE) microspheres were prepared through EHD co-jetting. Two different PLGA solutions with a concentration of 9 w/v% were prepared in solvent mixtures of chloroform and DMF (95:5, v/v), and P(HZ-*co*-BnGE) (0.9 w/v%, 10% by weight of PLGA) was introduced in one of the solutions. The polymer solutions were flown through side-by-side capillaries with the application of an appropriate voltage, and microspheres with a diameter of 2–3 μm were produced. The obtained microspheres were then collected from the jetting collector and the presence of hydrazide groups on the particle surface was confirmed via the conjugation of a fluorescent dye (RB-PEG-COOH) to the particles (Figure 35). By using EDC-coupling, the carboxylic acid groups of RB-PEG-COOH were activated to further react with the hydrazide groups on the microspheres. In CLSM images of the particles after the reaction, the red fluorescence layer from the rhodamine was observed only on the compartment containing P(HZ-*co*-BnGE) (green), which verified that the hydrazide groups were only on the surface of one compartment. Significantly, this demonstrated the selective localization of P(HZ-*co*-BnGE) to the pre-defined structures. Moreover, exploitation of the hydrazide groups of P(HZ-*co*-BnGE) to further modify the microsphere surface demonstrates potential reacting sites for biomolecules.

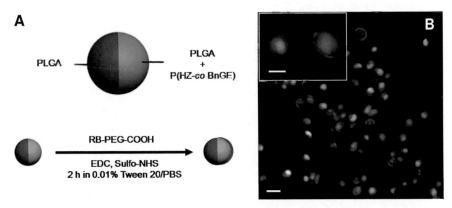

Figure 35   Modification of bicompartmental PLGA/P(HZ-*co*-BnGE) microparticles with RB-PEG-COOH (A) and CLSM images of these microspheres (B) showing selective binding of RB-PEG-COOH to the compartment containing P(HZ-*co*-BnGE). Blue, green, red dyes indicate PLGA, P(HZ-*co*-BnGE), and RB-PEG-COOH, respectively. Scale bar: 5 μm (inset 2μm).

P(HZ-co-BnGE) Microcylinders and Sugar-Conjugation

To validate the concept of using these functional microstructures for binding biomolecules, bicompartmental microcylinders containing P(HZ-*co*-BnGE) in one compartment were prepared and carbohydrates were selectively immobilized on them. The microcylinders were produced through the EHD co-jetting method, which was similar to that of the microspheres, except that the concentration of PLGA and polymer P(HZ-*co*-BnGE) were increased to 30 w/v% and 15 w/v% (50% by weight of PLGA), respectively. In addition, a rotary collector was introduced during the jetting to acquire highly aligned bundles of microfibers, and these microfibers were further processed to create uniform microcylinders by a previously reported microsectioning technique.[59] The selective localization of hydrazide groups on the obtained PLGA/P(HZ-*co*-BnGE) microcylinders was determined by confocal Raman microscopy (Figure 36). Through subsequent confocal Raman image scans, the hydrazide band at 1600 cm^{-1} was observed, and the confocal image demonstrated that the corresponding spectra was found only in one compartment of the microcylinder.

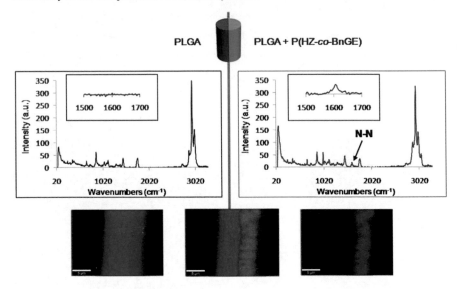

Figure 36   Confocal raman microscopy images of the bicompartmental PLGA/P(HZ-*co*-BnGE) microcylinders, showing the localization of P(HZ-*co*-BnGE) just in one compartment. Scale bars are all 5 μm.

Finally, the binding affinity of carbohydrates on the PLGA/P(HZ-*co*-BnGE) microcylinders was examined with 2α-mannobiose serving as a model carbohydrate. In order to bind 2α-mannobiose to the microcylinders, 2α-mannobiose was first oxidized through the addition of sodium periodate. The oxidation step was required to obtain free aldehyde groups, which was necessary for the covalent binding of 2α-mannobiose to the hydrazide groups on the cylinder surface. Furthermore, the successful binding of 2α-mannobiose to the microcylinders was unambiguously confirmed by introducing carbohydrate-lectin interactions. Lectins are well-known carbohydrate recognizing proteins, which show reversible and site-specific binding behavior, and thus, the presence of 2α-mannobiose on the microcylinders was determined by adding mannose-specific lectin, concanavalin A (Con A, rhodamine-labeled). After three hours of incubation with the 2α-mannobiose-bound microcylinders with Con A, the CLSM images of the microcylinders verified the selective binding of 2α-mannobiose to the bicompartmental PLGA/P(HZ-*co*-BnGE) microcylinders (Figure 37).

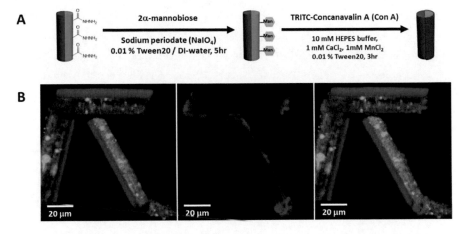

Figure 37   Reaction scheme (A) and CLSM images (B) showing selective binding of 2α-mannobiose followed by ConA conjugation on the bicompartmental PLGA/P(HZ-*co*-BnGE) microcylinders. The overlaid CLSM images demonstrate successful immobilization of 2α-mannobiose and ConA selectively on the P(HZ-*co*-BnGE) compartment. Blue, green, red dyes indicate PLGA, P(HZ-*co*-BnGE), and ConA, respectively.

4.2.4 Multifunctional *o*-Nitrobenzyl PEG and Photodegradable Particles

The materials in this chapter have been adapted with minor modifications from the following paper: E. Sokolovskaya, S. Rahmani, A. C. Misra, S. Bräse, J. Lahann, **2013**, submitted. *Synthesis of a Multifunctional Poly(ethylene glycol) Derivative and its Use in Dual-Stimuli Responsive Microparticles.*

4.2.4.1 Overview

Stimuli-Responsive Materials

Materials capable of changing their properties in a controlled and often reversible manner in response to their environment or external stimuli are defined as "smart" materials.[6] They can be designed either as switchable surfaces,[6,211,212] responsive hydrogels[7,189] or anisotropic particles.[6,8,9] Triggers include changes in temperature, pH, salt content, presence of specific molecules, electrical or magnetic fields, light and oxidative stress.[213,214,215] For example, thermoresponsive polymers have been used for preparation of shape memory materials.[216] Many strategies in the synthesis of these systems for controlled drug delivery rely on lower pH of endosomes or tumor tissues as compared to pH 7.4 of blood.[215,217] There are however very few examples of stimuli-responsive materials triggered by the oxidative environment common for inflammatory tissues.[215] Still inflammation relates to many diseases including neurodegenerative diseases (Parkinson's, Alzheimer's), atherosclerosis, inflammatory lung diseases and so on.[218] Therefore oxidative stress is an important trigger for controlled drug delivery.

Oxidative Stress Responsive Systems

Oxidative stress originates from the production of oxygen reactive species in inflammatory cells, starting from enzyme induced reduction of oxygen to superoxide anion followed by formation of hydrogen peroxide, hypochlorite, peroxynitrite and other reagents.[218] At low concentrations, oxidants are not harmful because of the presence of different enzymes in cells acting as antioxidants. In fact, oxidants even play a positive role since they activate protective mechanisms against pathogens. However at higher dosages, oxidants can cause strong damage to cells resulting in their death.[218] The presence of oxidants can be used as a marker of inflamed tissues and can activate a specific action of a biomaterial, such as drug release.

In one instance, hyaluronic acid based hydrogels were prepared for drug delivery system with controlled release of the drug upon degradation of hyaluronic acid in oxidative conditions.[219] Later another study, using natural polymer for drug delivery triggered by oxidative stress, was published by Frechet and co-workers.[220] They prepared nanoparticles from dextran modified with arylboronic esters in order to make the polymer water insoluble. In an oxidative environment, arylboronic esters groups are cleaved from dextran, thus allowing for fast release of the drug due to the dissolution of the particles.

Deeper study on oxidative stress reactive nanoparticles was performed by Hubbel and Tirelli,[221,222,223,224] who used poly(propylene sulfide) (PPS) as a polymer material. Thioethers are known to undergo facile oxidation into sulfoxides or sulfones upon treatment with hydrogen peroxide.[218] Due to the difference in hydrophilic properties of PPS and polysulfoxide or polysulfone – products of PPS oxidation, the former can be used for fabrication of drug delivery systems triggered by oxidative stress. This very promising system however lacks control over the degradation kinetics. Therefore this work has been extended by Mahmoud et al. by preparing nanoparticles from a polymer which combined both oxidants (thioether groups) and pH (acetal groups) sensitive moieties.[225] Thus a tandem sequence of reactions, including first dissolution of the particles due to the oxidation of polysufide followed by degradation of the polymer due to acetal groups hydrolysis lead to release of a loaded protein.

The use of dual stimuli-responsive carriers allows for more precise control over drug release and increases the degradation rates. However in the work of Mahmoud mentioned above, both stimuli were internal and acted simultaneously upon reaching the target tissue by the carrier. It would be more desirable to combine internal oxidative stress stimulus with an external control over the particle degradation. This can be achieved by combining hydrogen peroxide reactive moieties with photolabile groups.

Photoresponsive Systems and Protecting Groups

Compared to many known triggers, light switching may be performed in milder conditions and in a more specific manner due to control over its intensity and wavelength. Many photoresponsive systems have been developed thus far.[7,213,214] For example, azobenzene derivatives and spiropyrans were successfully used for the fabrication of photoresponsive surfaces.[212] In the field of drug delivery, photocleavable protecting groups are commonly used to switch the solubility of the polymer. For example, a block copolymer with a light-sensitive block was used for the generation of photodegradable micelle.[226] Polymer coatings suitable for photopatterning have potential use for creation of biological sensing systems as well as surfaces with phototunable wettability.

A variety of photolabile protecting groups for different functionalities have been developed thus far (Chapter 4.2.2.1). Among them *o*-nitrobenzyl (NB) group is the most commonly used.[227] Originally designed for use in organic synthesis[228] it was also successfully applied to biochemistry.[193] Photocleavage of NB group takes place readily upon irradiation with light of wavelengths higher than 300 nm releasing alcohol or acid and nitroso derivative (Scheme 33).[227]

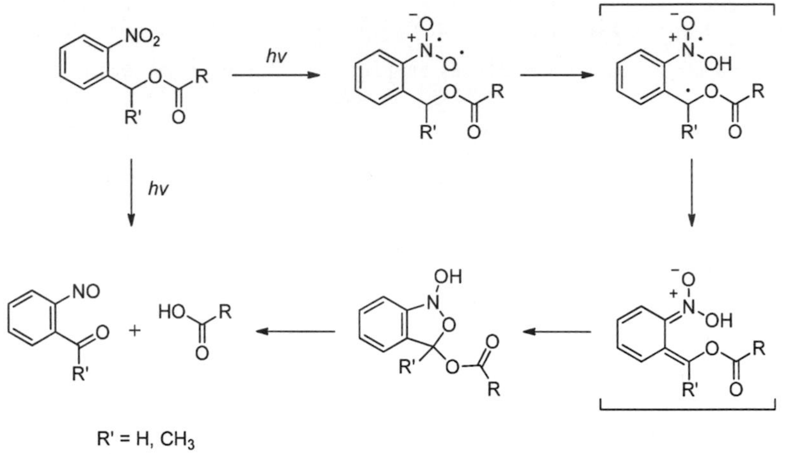

R' = H, CH₃

Scheme 33   Mechanism of photoinduced cleavage of NB protecting group.[227]

Dual Stimuli-Responsive Drug Carriers

Both oxidation of thioether into sulfoxide under conditions of an oxidative stress and photocleavage of NB protection with release of alcohol or acid moieties should result in increase of the polymer solubility. However the polymer can be designed such a way, that rapid degradation takes place just upon action of both stimuli, thus even if the polymer reached the oxidative environment no degradation is possible before UV irradiation, which should also be true for the reverse sequence of stimuli actions.

Here the development of new multifunctional *o*-nitrobenzyl PEG (NBPEG) suitable for fabrication of drug carriers with dually triggered release by oxidative stress and UV irradiation is presented.

4.2.4.2   Synthesis of Multifunctional NBPEG

Direct Polymerization

The synthesis of the target polymer containing both thioether and NB moieties can theoretically be performed by copolymerization of propylene sulfide, previously used by Hubbel and Tirelli,[221] and functional epoxide bearing NB protected group.

To access the feasibility of introducing NB protected moiety into PEG chain by direct polymerization, epoxide **49** was synthesized via two steps procedure as shown in Scheme 34.

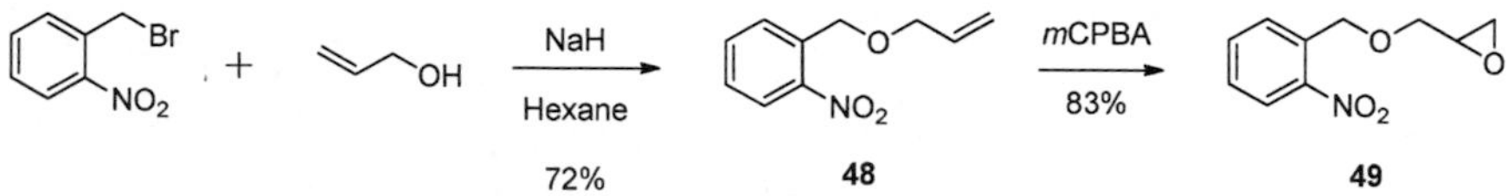

Scheme 34   Synthesis of the epoxide with NB protecting group.

Anionic ring-opening polymerization of the obtained epoxide **49** was further performed in the conditioned developed for the polymerization of other functional epoxides (Chapter 4.1.4.2.3). However, the polymerization was not successful and did not yield the desired polymer, because of the instability of the NB groups under the polymerization conditions.

Post-polymerization Modification

Since no suitable protection for NB groups is available, the only way to introduce them into the PEG chain is a post-polymerization modification approach. Still this method can be advantageous, because a single modification step can simultaneously introduce thioether groups and NB protection when thiol-ene reaction is applied. Previously this reaction has already been used as the first step in PHZ synthesis (Chapter 4.2.2.6), yielding a polyester with thioether linkage of ester groups to the PEG backbone. Here the same method has been applied for the introduction of NB moieties (Scheme 35).

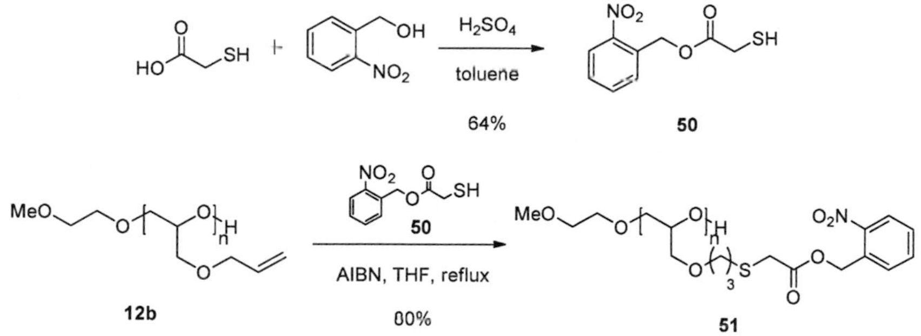

Scheme 35   Synthesis of NBPEG (**51**) by post-polymerization modification of PAGE via thiol-ene chemistry.

NB protected acid bearing thiol group **50** suitable for further thiol-ene modification has been synthesized by esterification of thioglycolic acid with *o*-nitrobenzyl alcohol. After purification on silica gel, thiol **50** has been used for the reaction with PAGE (**12b**) obtained previously (Chapter 4.1.4.2.2) to yield NBPEG (**51**). Full conversion of allyl moieties was confirmed by ^{1}H NMR spectrometry (Figures 38A and B). Polymer **51** was purified chromatographically on silica gel.

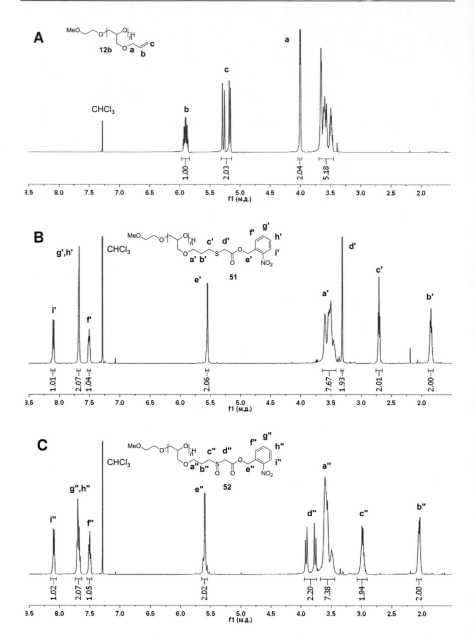

Figure 38   Comparative ^{1}H NMR spectra of polymers **12b** (A), **51** (B) and **52** (C). Solvent CDCl$_3$.

4.2.4.3 Oxidation of Thioether into Sulfoxide Groups in NBPEG

Under conditions of oxidative stress conversion of thioether groups of NBPEG into sulfoxide groups should take place, yielding more polar polymer. The latter should dissolve in water upon UV irradiation due to the cleavage of NB groups and release of the corresponding polyacid (Chapter 4.2.4.1). To compare the properties of NBPEG and its sulfoxide analogue, oxidation of NBPEG (**51**) with hydrogen peroxide into corresponding oxidized NBPEG (ONBPEG, **52**) has been performed (Scheme 36).

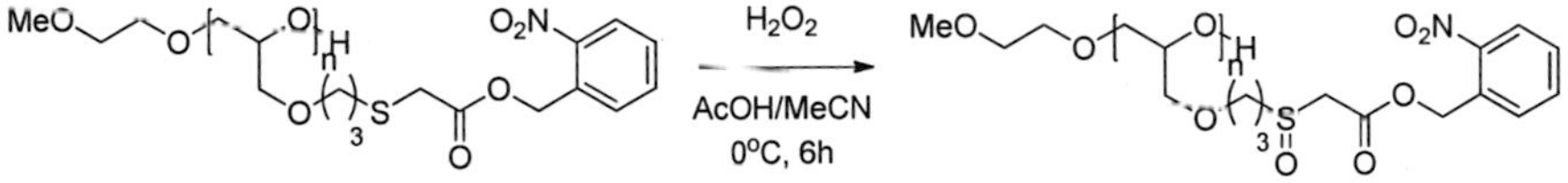

51   **52**

Scheme 36   Oxidation of polymer **51** into corresponding polymer **52**.

To ensure that oxidation of thioether moieties proceeds just till the formation of sulfoxide groups and no sulfone groups present in the product polymer, reaction was performed at 0°C. Oxidation to sulfone is undesirable not only because of the difference in solubility of sulfoxides and sulfones, but also because of the low stability of sulfoxides, which can undergo degradation upon irradiation.[218] Characterization of the obtained polymer **52** with ^{1}H NMR and IR spectrometry revealed no presence of sulfone moieties (Figures 38C, 39A). Instead complete oxidation of thioether groups was confirmed by disappearance of a signal at 3.28 ppm corresponding to the sulfur neighboring methylene protons d' and appearance of a new signal at 3.81 ppm corresponding to the same protons (d'') in polymer **52** (Figures 38B and C). This was also confirmed by IR spectra revealing appearance of a characteristic band at 1010 cm^{-1} corresponding to S=O stretching in sulfoxides (Figure 39A). GPC analysis confirmed that no degradation of polymer during oxidation takes place; instead a small shift of polymer **52** peak into the area of shorter elution times as compared to polymer **51** correlates with slight increase of molecular weight (Figure 39B).

A   **B**

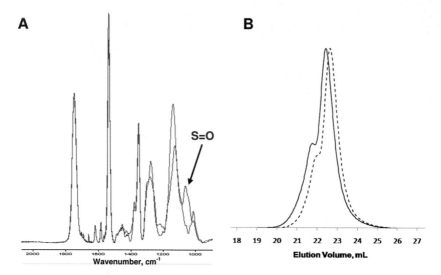

S=O

Wavenumber, cm⁻¹

Elution Volume, mL

Figure 39   Comparative IR spectra of polymers **51** (blue line) and **52** (red line) (A) and
their gel permeation chromatograms (polymer **51** (dashed line), polymer **52**
(solid line)) (B).

Since concentration of hydrogen peroxide in the conditions of oxidative stress *in vivo* is
relatively low, it has been suggested that *in vivo* oxidation of polysulfides results
predominantly in sulfoxides and not sulfones.[218] Therefore oxidation of NBPEG *in vivo*
should most probably lead to the formation of ONBPEG identical to those obtained in
laboratory conditions.

4.2.4.4   Photocleavage of NB Protection

Cleavage of NB protecting group in ONBPEG (**52**) should result in the formation of the
corresponding polyacid **53** (Scheme 37).

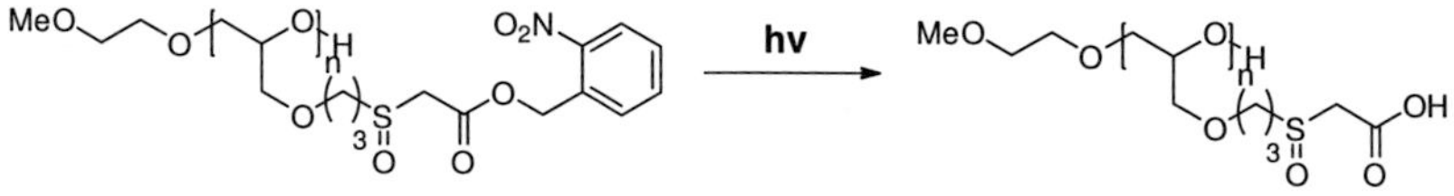

52   **53**

Scheme 37   UV induced cleavage of NB protecting groups in ONBPEG.

As a proof of concept deprotection of ONBPEG in DMSO-d_6 solution upon irradiation with 312 nm light was monitored by ^{1}H NMR spectrometry. Disappearance of signals corresponding to aromatic (e–h) and benzyl (d) protons of NB group explicitly confirms cleavage of NB protection (Figures 40). On the other hand methylene protons of the functional group (a–c) are still present in the spectrum with corresponding shifts (a'–c') after irradiation proving that no degradation of the polymer takes place (Figure 40). Dependence of the deprotection degree on irradiation time is shown in Figure 41.

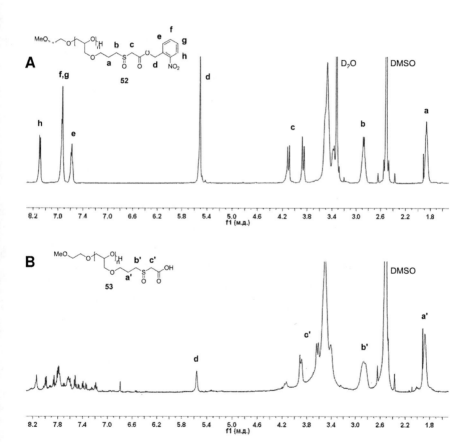

Figure 40   Comparative ^{1}H NMR spectra of ONBPEG (**52**) before (A) and after (B) 4 h of irradiation with 312 nm light. Solvent DMSO-d_6.

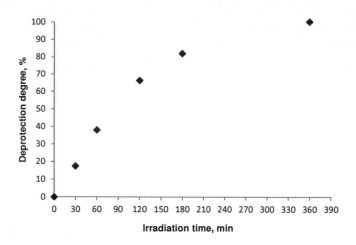

Figure 41   Dependence of ONBPEG deprotection degree on the time of irradiation with 312 nm light. Calculated based on ^{1}H NMR spectra.

Quite a long time (more than 4 h) is needed for complete cleavage of NB groups depending on the used solvent (DMSO) and quite high concentration of the polymer in the illuminated solution (5 mg/mL). These conditions were necessary for proper detection and assignment of signal in NMR spectra.

To define the time of polymer exposition to UV light for further experiments, ONBPEG was spin coated on gold wafers and exposed to 365 nm light for different periods of time. Cleavage of NB groups was monitored by IR spectrometry (Figure 42). Deprotection of NB groups was controlled by disappearance of characteristic nitro group bands at 1530 and 1346 cm^{-1}, as well as by shift of carbonyl group band from 1742 to 1729 cm^{-1}. Complete cleavage of NB protection was detected after irradiation for 30 min.

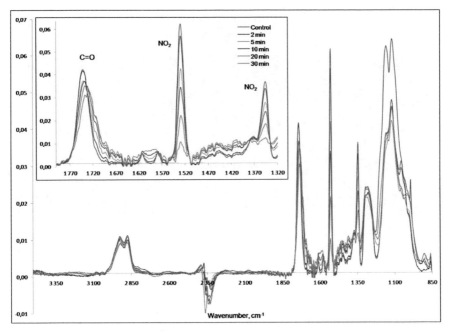

Figure 42   Comparative IR spectra of ONBPEG before (control) and after different times of irradiation with 365 nm light. Polymer was spin coated on the gold wafers.

4.2.4.5   Fabrication and Degradation of NBPEG/ONBPEG Microparticles

Fabrication of NBPEG/ONBPEG Microspheres

Anisotropic bicompartmental microspheres bearing photoresponsive ONBPEG in one compartment were generated by electrohydrodynamic co-jetting similar to the generation of those containing P(HZ-*co*-BnGE) (Chapter 4.2.3.3), except that the concentrations of PLGA and ONBPEG in the jetting solutions were changed. The total concentration of polymers in each solution was 15 w/v%, one solution contained PLGA only, another – PLGA/ONBPEG mixture in 1:1 w/w ratio. Both jetting solutions were loaded with fluorescent dyes, which allowed confirmation of the bicompartmental structure of the microspheres by localization of each dye in just one compartment and was observed in CLSM images of the generated particles (Figure 43).

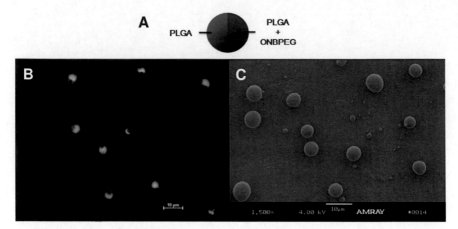

Figure 43   Schematic representation (A), CLSM (B) and SEM (C) images of bicompartamental microparticles with ONBPEG localized in one compatment. Green and blue dyes indicate PLGA and ONBPEG, respectively.

Furthermore, the localization of ONBPEG in just one compartment was additionally verified by confocal Raman microscopy (Figure 44). The characteristic band of the nitro group at 1348 cm^{-1} was used as a reference to identify the location of the polymer within a particle. Confocal Raman image scanning revealed the presence of this band and correspondingly ONBPEG in only one compartment of a microsphere.

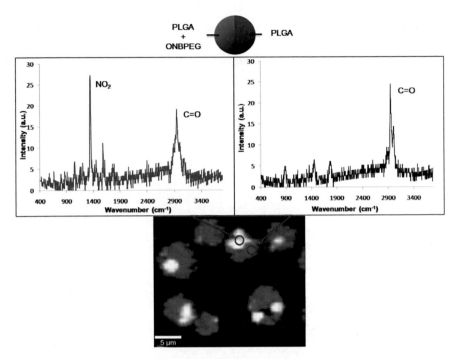

Figure 44   Confocal Raman microscopy images of the bicompartmental PLGA/ONBPEG microspheres, showing the localization of ONBPEG just in one compartment (the ratio PLGA:ONBPEG in one compartment was 3:1). Scale bar is 5 µm.

PLGA/NBPEG microspheres were produced using the same procedure as described above and were characterized with SEM and CLSM. The latter confirmed the bicompartmental structure of the obtained microspheres.

Degradation of NBPEG/ONBPEG Microspheres

The purpose of this study was to evaluate and compare the degradation ability of PLGA/NBPEG and PLGA/ONBPEG particles upon UV irradiation. In the conditions of oxidative stress, thioether moieties of NBPEG should undergo oxidation into sulfoxide groups yielding ONBPEG. The hydrophilicity of the latter is higher than that of NBPEG. However due to the presence of hydrophobic NB groups the solubility of both polymers should be insufficient for particle degradation. Photocleavage of NB protection in NBPEG and ONBPEG yields polyacids with thioether or sulfoxide moieties correspondingly (Figure 45A). Solubility of a polyacid should be higher than that of a protected polymer. However, due to the presence of hydrophobic thioether linkage the solubility of the polyacid obtained by deprotection of NBPEG should be still insufficient, while in case of ONBPEG deprotection should result in its rapid dissolution and, as a consequence, degradation of PLGA/ONBPEG particles is expected.

To verify this assumption, dry particles were exposed to 365 nm light for 30 min, after which they were incubated in TBS buffer for 1 h, washed with DI-water and analyzed by SEM. As a control experiment, incubation of non-irradiated particles in the same conditions was performed. The SEM images of irradiated particles and control experiments are shown in Figure 45.

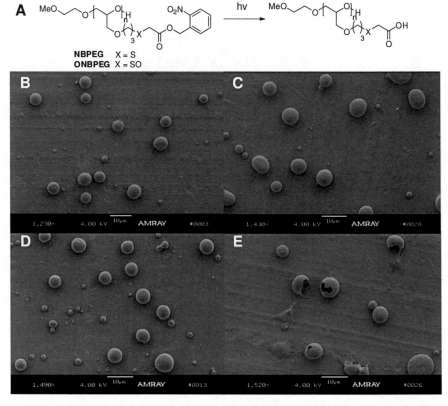

Figure 45   Deprotection of NBPEG/ONBPEG into the corresponding acid upon irradiation with the light of 365 nm (A) and SEM images of non-irradiated bicompartmental microparticles (PLGA/NBPEG (B) and PLGA/ONBPEG (C)) and after 30 min of irradiation with the 365nm light (PLGA/NBPEG (D) and PLGA/ONBPEG (E)), showing degradation of ONBPEG containing particles only after irradiation.

As expected no degradation was detected without irradiation of both PLGA/NBPEG and PLGA/ONBPEG particles (Figure 45B and C). Also the irradiated PLGA/NBPEG particles were stable in TBS buffer (Figure 45D). However, incubation of irradiated PLGA/ONBPEG particles led to fast dissolution of ONBPEG containing compartment and particles degradation, which was detected by SEM (Figure 45E). This confirms that corresponding PLGA/NBPEG particles are promising carriers for drug delivery with controlled release triggered by dual action of oxidants and UV irradiation.

ONBPEG Microfibers

To verify the potential to produce other shapes with ONBPEG and thus expand the utility of this polymer, biphasic fibers containing this polymer in one compartment were generated using the procedure described above for the generation of PLGA/P(HZ-*co*-BnGE) microfibers (Chapter 4.2.3.3).

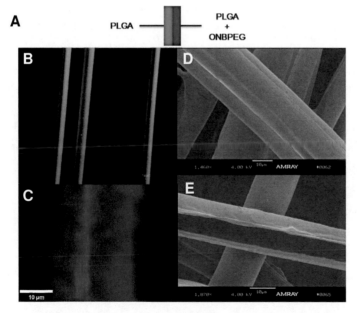

Figure 46   Schematic representation (A), CLSM (B), confocal Raman microscopy (C) and SEM (before (D) and after (E) 30 min of irradiation with 365 nm light) images of bicompartmental PLGA/ONBPEG microfibers with ONBPEG localized in one compartment. In CLSM image green and blue dyes indicate PLGA and ONBPEG, respectively (B). In confocal Raman microscopy image blue indicates presence of nitro groups and corresponds to the ONBPEG containing compartment, red indicates C=O and corresponds to PLGA compartment (C).

The fibers clearly have biphasic structure (Figure 46B) with ONBPEG localized in just one compartment as explicitly confirmed by confocal Raman microscopy (Figure 46C). Incubation of the fibers in TBS buffer in the same conditions as used before for the microspheres incubation, resulted in complete degradation of the ONBPEG containing compartment in case when the fibers were first irradiated with UV light, while no degradation was detected without UV irradiation.

The ability to produce fibers with ONBPEG has a potential application for the synthesis of degradable scaffolds for tissue engineering.

4.2.4.6   Cellular Uptake of ONBPEG Particles

Preparation of the particles for the in vitro studies and cell culture work for the Raw264.7 was completed by Sahar Rahmani and Asish Misra.

The *in vitro* studies to examine the ability of Raw264.7 cells to uptake PLGA/ONBPEG microparticles were performed. Particles containing green and red dyes in separate compartments were fabricated according to the procedure described in the previous section (Chapter 4.2.4.5). The particles were dried in vacuum to remove all remaining solvents, and then collected in a buffered solution, counted and incubated with cells. Several solutions with different concentrations of the particles (50,000 particles/mL, 100,000 particles/mL and 500,000 particles/mL) were used for the incubation with the cells performed for 4 h. The cellular uptake of PLGA/ONBPEG particles was then analyzed via CLSM imaging and quantified (Figure 47).

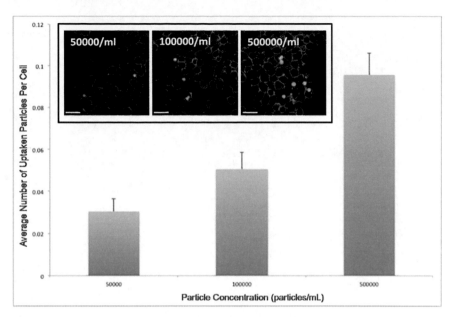

Figure 47   PLGA/ONBPEG microparticle uptake by Raw264.7 cells. Inset shows CLSM images of stained Raw264.7 cells after incubation with PLGA/ONBPEG microparticles at various concentrations (all scale bars are 20 µm). The graph shows the corresponding uptake quantified.

Because the Raw264.7 cells were derived from macrophages, they had the ability to phagocytose large particles as it can be seen in the CLSM images (inset, Figure 47). Both successful phagocytosis and frustrated phagocytosis were observed (Figure 48). This may be attributed to the variable size of the microparticles and inability of the cells to uptake the largest particles. Only particles completely uptaken were considered for the further quantification of the particle cellular uptake (Figure 47). An increase of the number of uptaken particles was observed at higher particle incubation concentrations (dose-dependence response), indicating nonspecific phagocytosis by Raw264.7 cells.

The performed cell study confirmed the suitability of PLGA/ONBPEG microparticles for drug delivery such as delivery to macrophages for potential immunotherapy applications.

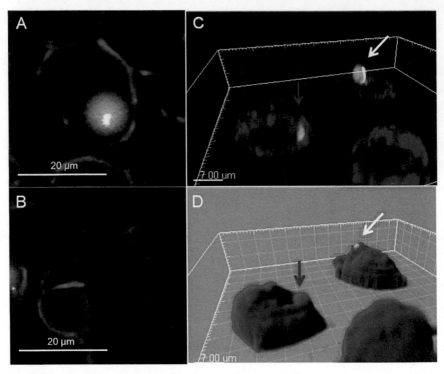

Figure 48   CLSM images of Raw264.7 cells showing the complete uptake (A) and the unsuccessful uptake ("frustrated phagocytosis") (B) of PLGA/ONBPEG microparticles and 3D CLSM renderings (using *Imaris* software) of the same z-stack: a transparent projection (C), and a shadow projection (D). The particle indicated by the red arrows is clearly uptaken, and therefore is not seen in the shadow projection rendering (D); the particles indicated by the yellow arrows is slightly protruding out of the cell in the shadow projection (D), indicating frustrated phagocytosis.

5 Summary and Perspectives

Two major aspects of the research in the area of multifunctional PEGs have been detailed in this work: preparation and investigation of the conditions for the synthesis of these polymers, and application of the obtained polymers for biomaterials fabrication.

5.1 Polymerization of EO and Functional Epoxides

In the first part of this thesis, a general method for the anionic ring-opening polymerization of EO and functional epoxides has been developed and employed for the synthesis of a number of multifunctional PEGs. In particular, different conditions for polymerization of EO and AGE as a representative functional epoxides have been employed and compared (Scheme 38). The best results were achieved, when polymerization was performed in THF solution (EO polymerization) or bulk (AGE polymerization) at temperatures below 30°C and with the use of PN as a deprotonating agent for *in situ* formation of an initiating alkoxide. Specifically in the case of AGE polymerization, thorough removal of oxygen traces from the polymerization media was found to be crucial in order to avoid crosslinking reactions of the double bonds, which leads to a significant broadening of the PAGE molecular weight distribution.

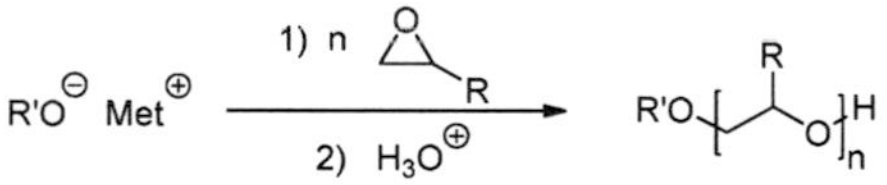

Scheme 38   General scheme for the anionic ring-opening polymerization of EO and functional epoxides.

Additional functionalization was realized for EO polymerization via initiation with functional alkoxides.

In the optimized conditions, polymerization of a number of functional epoxides with protected hydroxyl and aldehyde groups, has been performed. In all cases, conversions of the epoxides were high (79–99%) and polydispersities of the obtained polymers were low (PDIs: 1.06–1.09). Furthermore, di-, tri- and four arms star- block copolymers of EO and functional epoxides were prepared according to the developed procedure using mPEGs and PEGs as macroinitiators.

Next, the synthesized polymers were successfully purified chromatographically on silica gel. This easy and efficient technique has rarely been applied for polymer purification before and allows multifunctional PEGs free of small molecular weight impurities to be obtained. Moreover for copolymers, this method resulted in the removal of all traces of homopolymers present in the reaction.

In two instances: PEEGE and PGADEA, acetal protection of functional groups was employed during polymerization. Successful post-polymerization cleavage of the protecting groups in PEEGE upon treatment with acid was performed providing the corresponding PG in an excellent yield (75%). Unfortunately, deprotection of aldehyde groups in PGADEA proved to be difficult most likely due to the numerous Aldol reactions, which can take place in the presence of an acid. Alternatively, multifunctional PEGs with photolabile acetal protection for aldehyde groups have been prepared and showed reactivity toward hydrazides upon UV treatment.

5.2   Biomedical Applications of Multifunctional PEGs

In the second part of the thesis, efforts were focused on the fabrication of biomaterials, including stimuli-responsive materials, based on the prepared functional epoxides and multifunctional PEGs obtained directly by polymerization and by post-polymerization modification of PAGE.

Reactive surfaces were prepared by microcontact printing of an epoxide bearing activated alkyne group on the amino functionalized substrates and further modified with biotin-azide via copper-free alkyne-azide click reaction. Successful biotin immobilization was confirmed through subsequent conjugation of RB-streptavidin. However, unspecific binding to amino functionalized surfaces was observed, which indicated the necessity for additional passivation of non-functionalized areas after reaction with the epoxide.

Smart hydrogels were prepared from multifunctional responsive copolymer PPAPEG-PEG-PPAPEG and PHZ via light induced aldehyde-hydrazide click reaction. UV irradiation caused cleavage of the photosensitive acetal protection in the PPAPEG blocks, releasing free aldehyde moieties which readily reacted with the hydrazide groups of PHZ forming acid labile hydrazone linkages. This system allows for controlled gel formation under mild conditions of UV irradiation compatible with the loading of biomolecules or cellular encapsulation and opens the possibility for *in situ* gelation upon injection. Moreover, crosslinking via labile hydrazone linkages provides degradability of the gel at lower pH, which has a potential for drug delivery with controlled release.

PHZ, used as one of the components for hydrogel formation was prepared via two step modification of PAGE including a thiol-ene reaction as the first step. The hydrophilic properties of PHZ were tuned by introduction of hydrophobic benzyl moieties into the polymer chain. Due to its insolubility in water, the P(HZ-*co*-BnGE)) could be utilized for the fabrication of functional bicompartmental microparticles via EHD co-jetting. The prepared particles reacted with oxidized 2α-mannobiose via formation of hydrazone bonds specifically in the compartment containing P(HZ-*co*-BnGE), which was confirmed by subsequent successful carbohydrate-lectin interactions. Selectively sugar-modified particles can further contribute to the development of targeted drug and gene delivery based on sugar recognition by lectins.

Finally, stimuli-responsive degradable bicompartmental microparticles were prepared from NBPEG by EHD co-jetting. NBPEG was synthesized from PAGE via thiol-ene reaction with the corresponding thiol bearing photosensitive NB group. This polymer changes its solubility in water upon dual action of light and hydrogen peroxide. UV irradiation stimulated cleavage of NB groups while hydrogen peroxide oxidized thioether into sulfoxide groups. Both reactions increased the hydrophilicity of NBPEG, however only simultaneous occurrence of these transformations resulted in a fast dissolution of the polymer in water. This was confirmed by comparison of the degradation rates of the irradiated and non-irradiated particles containing NBPEG and its oxidized analogue – ONBPEG. Only irradiated ONBPEG particles showed rapid degradation of one compartment (containing functional polymer) in water solution. Furthermore, successful uptake of ONBPEG particles by macrophages Raw264.7 was shown. NBPEG particles are promising carriers for drug delivery with controlled release triggered by dual action of oxidative stress and UV irradiation.

5.3 Perspectives for the Future Development and Applications of Multifunctional PEGs

The performed work demonstrated that multifunctional PEGs are a highly efficient tool for material scientists with significant potential for biomedical applications. On-demand properties can be assigned to multifunctional PEGs depending on the introduced functionalities and the method of modification. This is especially valuable for the rapidly developing field of smart biomaterials since various new stimuli-responsive polymers can be prepared by incorporation of suitable moieties. Moreover, multifunctional PEGs are easily accessible via anionic ring-opening polymerization and, when needed, subsequent post-polymerization modification via click chemistries. The living nature of anionic ring-opening polymerization allows polymers with very narrow molecular weight distribution to be obtained and for the simple synthesis of block copolymers by subsequent addition of different monomer epoxides. Furthermore, this technique can even be combined with carbanionic polymerization due to recent advances in coordination polymerization.

Still this is a quite new niche in polymer science and has not yet been deeply investigated although intensive work is currently carried out towards its development. The following directions for the further research in this area can be considered:

1.   Broadening the library of epoxides with functional groups which are either resistant under the polymerization conditions or can be protected with a suitable protecting group stable under the polymerization conditions and easily cleavable thereafter. Functional groups suitable for further modification via click chemistries are particularly attractive.

      Conversely, this challenge can be also approached by optimization of polymerization conditions such as finding new initiators or additives that decrease the basicity of the polymerization environment. For example, coordination polymerization is an alternative to anionic polymerization in some instances.

2.   Development of the modification procedures of already known multifunctional PEGs. Thus far only PAGE serves as a basic material for the synthesis of new functional PEGs since easy and efficient modification via thiol-ene reaction is available. Modification of PG and PECH has also been reported, however, in contrast to PAGE, their modification involves ordinary chemical reactions often resulting in incomplete

conversion and uncontrolled presence of different moieties within one polymer molecule. Known multifunctional PEGs such as PGADEA and PFO or PPAPEG presented herein could be alternatives to PAGE since they are also suitable for click chemistry modification.

3.   Synthesis of multifunctional PEGs with new architectures. At this time, only PG has been employed for some applications as a hyperbranched structure. Branched architectures would give new properties to multifunctional PEGs different from those of the linear structures.

4.   Finally, investigation of the methods to process multifunctional PEGs is absolutely vital since they define the final application of the polymer. Multifunctional PEGs offer unique properties which can be utilized differently depending on the processing technique, such as surface modification, particles fabrication or hydrogel synthesis.

To conclude, synthesis and application of multifunctional PEGs is a rapidly developing and high-potential area of material science which opens new perspectives for biomaterials synthesis for diverse applications in biomedicine.

6   Experimental Section

6.1   Materials and Methods

6.1.1   Reagents and Solvents

Unless otherwise specified, all chemicals and solvents were purchased from commercial suppliers (*Sigma-Aldrich, Alfa Aesar, Merck KGAA, BDH Prolabo, Vector Laboratories, Thermo Scientific, Nanocs Inc., Life Technologies, Andwin Scientific*) and used without further purification.

Absolute THF was prepared by distillation from sodium/benzophenone under an argon atmosphere. Dry DMSO was prepared by distillation from calcium hydride under an argon atmosphere. Dry solvents were used immediately after preparation. All other dry solvents were commercial and used without additional preparation.

Methyl 2-hydroxy-5-methoxybenzoate (**29**) was prepared as described elsewhere.[229] Naphthalene was recrystallized in methanol. PN was prepared from recrystallized naphthalene and potassium in absolute THF according to procedure described elsewhere.[62] EO and all liquid initiating alcohols and epoxides were freshly distilled from calcium hydride under an argon atmosphere prior to polymerization. For some experiments AGE was additionally distilled from butyl magnesium chloride and degassed with the use of freeze-pump-thaw technique prior to polymerization. All solid initiating alcohols, epoxides and PEGs used as macroinitiators were dried by stirring in benzene at 60°C for 15–30 min, followed by evaporation of benzene *in vacuo* for at least 2 h at 60–70°C prior to polymerization.

Amino functionalized surfaces were prepared by CVD-process of amino paracyclophane on the silicon and gold wafer according to the previously reported procedure.[186]

6.1.2   Preparative Procedure

All moisture or air sensitive reactions and all polymerizations were carried out in flame-dried Schlenk-flasks under an argon atmosphere. Liquid reagents were added to the reaction vessel with plastic syringes or gastight syringes purchased from *Hamilton* or transferred via cannula. Powdered solids were added under an argon flow.

Unless otherwise specified, the reactions were carried out at room temperature. For the reactions performed at low temperatures the following cooling mixtures were used:

0°C   ice/water

–78°C   acetone/dry ice

Controlled of the reactions was done by thin layer chromatography and ^{1}H NMR spectroscopy.

Unless otherwise specified, all organic solvents were evaporated under reduced pressure at 40°C and water – at 50°C on a *Heidolph* Rotary Evaporator.

6.1.3 Product Purification

Small molecular weight compounds were purified by distillation, recrystallization or chromatographically on silica gel (Geduran Si 60, 0.040–0.063 mm, 230–400 mesh ASTM purchased from *Merck*). Polymers were purified by precipitation in cold diethyl ether, chromatographically on silica gel (Geduran Si 60, 0.040–0.063 mm, 230–400 mesh ASTM purchased from *Merck*) or by dialysis against Milli-Q (18.2 MΩcm) water with dialysis membrane Spectra/Por 6 (MWCO = 1000 g/mol) purchased from *Spectrum Laboratories*.

6.1.4 Characterization

6.1.4.1 Compounds Characterization

6.1.4.1.1 Nuclear Magnetic Resonance (NMR)

^{1}H NMR and ^{13}C NMR spectra were recorded on *Bruker* Avance III 500 spectrometer at room temperature with deuterated solvents (chloroform-d_1, DMSO-d_6, acetonitrile-d_3, methanol-d_4, water-d_2) purchased from *Eurisotop*. Working frequencies were 500 MHz for ^{1}H and 125 MHz for ^{13}C NMR.

Chemical shifts δ were given in parts per million (ppm) and referenced to chloroform (^{1}H: δ = 7.26 ppm, ^{13}C: δ = 77.16 ppm), DMSO (^{1}H: δ = 2.50 ppm, ^{13}C: δ = 39.52 ppm), acetonitrile

(^{1}H: δ = 1.94 ppm, ^{13}C: δ = 1.32 ppm), methanol (^{1}H: δ = 3.31 ppm, ^{13}C: δ = 49.00 ppm).[230] Coupling constants (*J*) were given in Hertz (Hz): s = singlet, br s = broad singlet, d = doublet, t = triplet, q = quartet, m = multiplet, dd = doublet of doublets, dt = doublet of triplets, ddd = doublet of dd, ddt = doublet of dt, td = triplet of doublets, tt = triplet of triplets. The following abbreviations for signals were used: H_{Ar} = aromatic proton, C_{Ar} = aromatic carbon, H_{epoxy} = epoxide ring protons, C_{epoxy} = epoxide ring carbons, H_{PEG} = PEG backbone protons, C_{PEG} = PEG backbone carbons. The signals structure in ^{13}C NMR spectra was analyzed by DEPT-technique (DEPT = Distortionless Enhancement by Polarization Transfer) and given as following: C = quaternary carbon (no DEPT-signal), CH = tertiary carbon (positive DEPT-signal), CH_2 = secondary carbon (negative DEPT-signal), CH_3 = primary carbon (positive DEPT-signal).

6.1.4.1.2 Infrared (IR) Spectroscopy

IR spectra were recorded on *Bruker* IFS 88 or *Bruker* Alpha spectrometers. The samples were measured as thin films between KBr plates (oils, liquids) or as pure compounds using ATR-technique (ATR = Attenuated Total Reflection) (solids). The position of the absorption band is given in wavenumbers (υ) in cm^{-1} in an interval between 3600 cm^{-1} and 500 cm^{-1}. The intensities of the bands were characterized as follows: vs = very strong (0–10% T), s = strong (10–40% T), m = medium (40–70% T), w = weak (70–90% T), vw = very weak (90–100% T); T = Transmission.

Analysis of the polymer structure was also undertaken with a *Thermo* Nicolet 6700 spectrometer with an 85° grazing angle and 128 scans for each sample at a resolution of 4 cm^{-1} or with *Bruker* Vertex 80 spectrometer, detector: LN2 cooled narrow band MCT with an 80° grazing angle and 1024 scans for each sample at a resolution of 2 cm^{-1}. Reference: deuterated 1-hexadecanethiol SAM on gold wafers. Samples were prepared by spin coating of polymer solutions (in dry THF or chloroform) on gold wafers.

6.1.4.1.3 Ultraviolet-Visible (UV-Vis) Spectroscopy

The absorption of polymers was determined on *Varian* Cary 50 Bio UV-Vis spectrophotometer using quartz cuvette (1 cm).

6.1.4.1.4 Mass Spectrometry (MS)

Electron ionization (EI) mass spectra were recorded on a *Finnigan* MAT 95 spectrometer. The molecular fragments are quoted as a relation between mass and charge (m/z), the intensities as a percentage value relative to the intensity of the base signal (100%). The abbreviation $[M]^+$ refers to the molecular ion.

For high resolution measurements the following abbreviations were used: calcd. = theoretical value, found = measured value.

Matrix-assisted laser desorption/ionization time of flight (MALDI-TOF) mass spectra were recorded on an *AB SCIEX* 4800 Proteomics Analyzer. α-Cyano-4-hydroxycinnamic acid and dithranol were used as the matrixes and sodium iodide as an ionizing agent.

6.1.4.1.5 Elemental Analysis (EA)

Elemental analysis was performed on *Elementar* vario MICRO analyzer. The samples were weighed with *Sartorius* M2P analytical balance. The amount of carbon (C), hydrogen (H), nitrogen (N) and sulfur (S) is quoted as mass percent. The following abbreviations were used: calcd. = theoretical value, found = measured value.

6.1.4.1.6 Gel Permeation Chromatography (GPC)

GPC in THF was performed on an *Toson* EcoSEC GPC system with a differential refractive index detector, using THF as mobile phase at 30°C with an elution rate of 1.0 mL/min. Three *PSS* SDV columns (100 Å, 5 µ, 8.0 × 300 mm, 1000 Å, 5 µ, 8.0 × 300 mm and 100000 Å, 5 µ, 8.0 × 300 mm) were calibrated by linear PS standards (*PSS*).

GPC in N,N-dimethylacetamide (DMAc) was performed on *Polymer Laboratories* PL-GPC 50 Plus Integrated System with autosampler and a differential refractive index detector, using DMAc with 0.03 w/w% LiBr as mobile phase at 50°C with an elution rate of 1.0 mL/min. A *Polymer Laboratories* PLgel 5 µm bead-size guard column (7.5 × 50 mm) and three *Polymer Laboratories* PLgel 5 µm MixedC columns (7.5 × 300 mm) were calibrated by linear PS standards (*PSS*).

6.1.4.1.7 Thin layer chromatography (TLC)

TLC was performed on silica gel plates (TLC Silica gel 60 F_{254}, layer thickness 0.25 mm, purchased from *Merck*) followed by detection with ultraviolet light (*UVP* handheld UV lamp UVGL-58, $\lambda = 254$ nm) or with the following developing solutions:

- potassium permanganate solution (1.0 g of $KMnO_4$ and few drops of conc. H_2SO_4 in 200 mL water)

- Seebach solution (2.5 g of phosphomolybdic acid, 1.0 g of cerium(IV)sulfate tetrahydrate in 6 mL of conc. H_2SO_4 und 90.5 mL H_2O)

- Dragendorf mixture (mixture of solutions A and B diluted to 100 mL with H_2O; solution A: 0.17 g of bismuth nitrate in 2 mL of AcOH and 8 mL of H_2O; solution B: 4 g of KI in 10 mL of AcOH and 20 mL of H_2O) for PEG derivatives.

6.1.4.1.8 Melting points

Melting points were measured on a *Laboratory Devices Inc.*, Model Mel-Temp II melting point microscope and were not corrected.

6.1.4.2 Surface Analysis

6.1.4.2.1 X-ray Photoelectron Spectroscopy (XPS) Analysis

XPS was used to assess the composition of the compounds on the surface before and after modification. Specifically, data were recorded with an Axis Ultra X-ray photoelectron spectrometer (*Kratos Analyticals*) outfitted with a monochromatized Al Ka X-ray source at a power of 150 kW. Survey spectra were taken at 160 eV.

6.1.4.2.2 Fluorescence Microscopy

Imaging of the surfaces modified by microcontact printing was performed with *Zeiss* Axioplan 2 imaging system and Zeiss AxioCamMRm camera. For red fluorescence the following set of filters was used: BP 546/12, FT 580, LP 590.

6.1.4.3 Particles Analysis

6.1.4.3.1 Confocal Laser Scanning Microscopy (CLSM)

Particles were deposited on glass cover slips during the jetting process. A drop of water was placed on the cover slip and the particles were visualized using a CLSM *Olympus* FluoView 500. 405 nm laser, 488 nm Argon laser, and 533 nm Helium-Neon green (HeNeG) laser were used to excite MEHPV, PTDPV and TRITC (labeling dyes for Con A), respectively. The barrier filters were set to 430–460 nm for MEHPV, 505–525 nm for PTDPV, and >560nm for TRITC.

6.1.4.3.2 Scanning Electron Microscopy (SEM)

Particles suspended in water were deposited on silicon wafers and left to dry overnight. The wafers were sputter coated with gold and then imaged via an *Amray* 1910 Field Emission Scanning Electron Microscope (FEG-SEM).

6.1.4.3.3 Confocal Raman Microscopy

The confocal Raman microscopy was performed on *WITec* alpha 300R utilizing 532 nm laser. Microparticles and microfibers were dried under vacuum overnight prior to measurements to remove any remaining solvents. Spectra were acquired using an integration time of 0.5 sec per pixel with image scan area of 100 pixels × 100 pixels.

6.2 Synthesis and Characterization of Small Molecular Weight Compounds

6.2.1 Initiating Functional Alcohols

2-Azidoethanol (**2e**)[231] and 4-(2-hydroxyethyl)-10-oxa-4-azatricyclo[5.2.1.02,6]dec-8-ene-3,5-dione (**2f**)[232] were prepared according to previously reported procedures.

N-(2-Hydroxyethyl)-9-anthracenecarboxamide (2c)

1 g (4.5 mmol, 1 eq) of 9-anthracenecarboxylic acid and 0.69 g (4.5 mmol, 1 eq) of 1-hydroxybenzotriazole hydrate (BtOH·H$_2$O) were dissolved in 40 mL of dichloromethane and cooled down to 0°C. N,N'-Diisopropylcarbodiimide (DIC) (0.76 mL, 4.95 mmol, 1.1 eq) was added dropwise and the reaction was stirred at 0°C for 4 h. Formed precipitate was filtered off, ethanolamine (1.36 mL, 22.5 mmol, 5 eq) was added dropwise to the cooled down to 0°C filtrate and reaction was stirred overnight at room temperature. The reaction was quenched with water, the organic phase was separated and washes with 0.1 M HCl, water, 0.1 M NaOH, water and dried over anhydrous sodium sulfate. Solvent was removed *in vacuo* and the crude product was recrystallized from ethyl acetate to yield 0.63 g (53%) of the title compound as a light beige solid. Mp 205–208°C (decomposition). ^{1}H NMR (500 MHz, CDCl$_3$): δ = 8.47 (s, 1H, H$_{Ar}$), 8.08 (d, J = 8.7 Hz, 2H, H$_{Ar}$), 8.00 (d, J = 8.3 Hz, 2H, H$_{Ar}$), 7.54–7.46 (m, 4H, H$_{Ar}$), 6.49 (br s, 1H, NH), 3.98 (t, J = 5.0 Hz, 2H, C$\underline{H}_2$OH), 3.86–3.82 (m, 2H, C$\underline{H}_2$NH), 2.35 (br s, 1H, OH) ppm; ^{13}C NMR (125 MHz, CDCl$_3$): δ = 170.9 (C=O), 131.2 (C$_{Ar}$), 128.7 (2C, CH$_{Ar}$), 128.6 (CH$_{Ar}$), 128.4 (2C, C$_{Ar}$), 128.2 (2C, C$_{Ar}$), 127.0 (2C, CH$_{Ar}$), 125.7 (2C, CH$_{Ar}$), 125.1 (2C, CH$_{Ar}$), 62.8 (CH$_2$OH), 43.1 (CH$_2$NH) ppm; FTIR (Platinum ATR): υ = 3245 (vw), 3051 (vw), 2849 (vw), 2745 (vw), 1933 (vw), 1682 (w), 1627 (w), 1576 (w), 1435 (w), 1387 (w), 1296 (w), 1260 (w), 1200 (w), 1159 (w), 1107 (w), 1060 (w), 1039 (w), 1014 (w), 899 (vw), 874 (vw), 821 (vw), 790 (vw), 728 (w), 671 (w), 628 (w), 601 (w), 557 (vw), 537 (vw) cm^{-1}; MS (EI, 70eV), m/z (%): 265.1 (100) [M]$^+$, 221.1 (32) [C$_{15}$H$_{11}$NO]$^+$, 205.0 (79) [C$_{15}$H$_9$O]$^+$, 177.0 (85) [C$_{14}$H$_9$]$^+$; HRMS calcd. for C$_{17}$H$_{15}$NO$_2$: 265.1103; found 265.1101.

6.2.2   Functional Epoxides

2-(3-Trimethylsilyl-2-propynyl)oxirane (6b)

The synthetic procedure was adapted from those which was previously reported.[150,151] 12.3 mL (86.8 mmol, 1.5 eq) of trimethylsilylacetylene were dissolved in 230 mL of hexane under an argon atmosphere and cooled down to 0°C. Next, 36.2 mL (1.6 M, 58.0 mmol, 1 eq) of n-BuLi solution was added and the reaction was stirred for 10 min at 0°C, followed by addition of diethylaluminum chloride (1.0 M, 58 mL, 58.0 mmol, 1 eq). The reaction was stirred at 0°C for an additional 30 min, after which epichlorohydrin (4.5 mL, 58.0 mmol, 1 eq) was introduced dropwise. The reaction was stirred at 0°C for 2.5 h and quenched with 1 M HCl. The product was extracted with diethyl ether, washed with brine, dried over anhydrous sodium sulfate, filtered and concentrated *in vacuo* yielding 9.2 g (83%) of chlorohydrin as a colorless liquid, which was used for the next step without purification.

Crude chlorohydrin **6a** (9.2 g, 48.3 mmol, 1 eq) was dissolved in dichloromethane (100 mL) followed by slow introduction of sodium hydroxyl anion (2.9 g, 72.4 mmol, 1.5 eq). The reaction was stirred for 2 days and quenched with saturated ammonium chloride solution. The mixture was extracted with dichloromethane, and the combined organic extracts were washed with brine, dried over anhydrous magnesium sulfate, filtered and concentrated *in vacuo*. The crude product was purified by distillation to yield 5 g (67%) of epoxide **6b** as a colorless liquid. Bp 65–66°C/11 mbar. ^{1}H NMR (500 MHz, CDCl$_3$): δ = 3.11–3.06 (m, 1H, CH$_{epoxy}$), 2.78 (dd, J = 4.7 Hz, J = 4.1 Hz, 1H, CH$_{2\text{-epoxy}}$), 2.68–2.62 (m, 2H, 1× CH$_{2\text{-epoxy}}$ + 1×CH$_2$C≡C), 2.47 (dd, J = 17.5 Hz, J = 5.2 Hz, 1H, CH$_2$C≡C), 0.14 (s, 9H, CH$_3$) ppm; ^{13}C NMR (125 MHz, CDCl$_3$): δ = 100.8 (C≡$\underline{C}$Si), 87.2 ($\underline{C}$≡CSi), 49.8 (CH$_{epoxy}$), 46.5 (CH$_{2\text{-epoxy}}$), 23.6 ($\underline{C}$H$_2$C≡C), 0.1 (3C, SiCH$_3$) ppm. Analytical data for epoxide **6b** matched literature data.[150]

N-Glycidylphthalimide (7)

The synthetic procedure was adapted from that which was previously reported.[154] Potassium phthalimide (12.95 g, 70 mmol, 1 eq) and benzyltriethylammonium chloride (1.59 g, 7 mmol, 0.1 eq) were suspended in 130 mL of *i*-propanol, and 16.4 mL (210 mmol, 3 eq) of epichlorohydrin were added thereto. The reaction stirred for 48 h at room temperature. *i*-Propanol was then removed *in vacuo*, the residue was dissolved in ethyl acetate, washed with water and dried over anhydrous anhydrous magnesium sulfate. Ethyl acetate was then removed *in vacuo*, and the crude product was recrystallized from ethyl acetate/hexane mixture to yield 10.16 g (67%) of epoxide **7** as a white solid. Mp 97–100°C. ^{1}H NMR (500 MHz, CDCl$_3$): δ = 7.88–7.83 (m, 2H, H$_{Ar}$), 7.75–7.70 (m, 2H, H$_{Ar}$), 3.96 (dd, J = 14.4 Hz, J = 5.1 Hz, 1H, CH$_2$N), 3.81 (dd, J = 14.4 Hz, J = 5.0 Hz, 1H, CH$_2$N), 3.26–3.22 (m, 1H, CH$_{epoxy}$), 2.81 (dd, J = 4.9 Hz, J = 4.2 Hz, 1H, CH$_{2\text{-epoxy}}$), 2.69 (dd, J = 4.9 Hz, J = 2.5 Hz, 1H, CH$_{2\text{-epoxy}}$) ppm; ^{13}C NMR (125 MHz, CDCl$_3$): δ = 168.1 (2C, C=O), 134.3 (2C, CH$_{Ar}$), 132.1 (2C, C$_{Ar}$), 123.6 (2C, CH$_{Ar}$), 49.2 (CH$_{epoxy}$), 46.2 (CH$_{2\text{-epoxy}}$), 39.8 (CH$_2$N) ppm. Analytical data for epoxide **7** matched literature data.[154,233]

1-Ethoxyethyl glycidyl ether (EEGE, 8)

The synthetic procedure was adapted from that which was previously reported.[156] 14.3 mL (216 mmol, 1 eq) of glycidol were dissolved in 80 mL (837 mmol, 3.9 eq) of ethyl vinyl ether followed by slow addition of *p*-toluenesulfonic acid monohydrate (0.4 g, 2.1 mmol, 0.01 eq) while maintaining the reaction temperature below 40°C. Reaction was stirred for 3 h at room temperature and quenched by addition of saturated NaHCO$_3$ solution. The organic phase was separated, dried over anhydrous magnesium sulfate, filtered and concentrated *in vacuo*. The crude product was distilled *in vacuo* to yield 16.9 g (54%) of epoxide **8** (inseparable mixture of diastereomers) as a colorless liquid. Bp 152–156°C. ^{1}H NMR (500 MHz, CDCl$_3$): δ = 4.75

and 4.73 (q, J = 5.4, 1H, C<u>H</u>CH$_3$), 3.81–3.37 (m, 4H, CH$_2$O + C<u>H</u>$_2$CH$_3$), 3.16–3.10 (m, 1H,

CH$_{epoxy}$), 2.78 and 2.78 (dd, J = 5.1, J = 4.1, 1H, CH$_{2\text{-}epoxy}$), 2.62 and 2.59 (dd, J = 5.1, J =

2.7, 1H, CH$_{2\text{-}epoxy}$),1.30 and 1.29 (d, J = 5.4 Hz, CHC<u>H</u>$_3$), 1.18 (t, J = 7.1, 3H, CH$_2$C<u>H</u>$_3$) ppm;

^{13}C NMR (125 MHz, CDCl$_3$): δ = 99.7 and 99.6 (CH$_3$<u>C</u>H), 65.7 and 65.1 (CH$_2$O), 60.9 and

60.8 (<u>C</u>H$_2$CH$_3$), 50.9 and 50.8 (CH$_{epoxy}$), 44.5 and 44.5 (CH$_{2\text{-}epoxy}$), 19.7 and 19.6 (<u>C</u>H$_3$CH),

15.2 (CH$_2$<u>C</u>H$_3$) ppm. Analytical data for epoxide **8** matched literature data.[156]

Glycidyl 2-bromo-2-methylpropanoate (9)

The synthetic procedure was adapted from that which was previously reported.[157] Glycidol

(3.45 mL, 52 mmol, 1.3 eq) and triethylamine (6.7 mL, 48 mmol, 1.2 eq) were dissolved in

200 mL of dichloromethane and cooled down to 0°C. Next, 2-bromo-2-methylpropanoyl

bromide (5.0 mL, 40 mmol, 1 eq) was added thereto dropwise and reaction was stirred for 1 h

at room temperature. Reaction was quenched with water, organic layer was separated, washed

sequentially with 10% citric acid, water, half saturated solution of sodium hydrogen

carbonate, water and dried over anhydrous magnesium sulfate. Solvent was evaporated *in*

vacuo and the crude product was purified by column chromatography on silica gel with

dichloromethane/ethyl acetate (9:1 v/v) mixture as an eluent to yield 7.96 g (89%) of the pure

epoxide **9** as a colorless liquid. R$_f$ = 0.97 .^{1}H NMR (500 MHz, CDCl$_3$): δ = 4.48 (dd, J = 12.3

Hz, J = 3.0 Hz, 1 H, OCH$_2$), 4.06 (dd, J = 12.3 Hz, J = 5.9 Hz, 1 H, OCH$_2$), 3.27–3.23 (m,

1H, CH$_{epoxy}$), 2.86 (dd, J = 4.9 Hz, J = 4.3 Hz, 1 H, CH$_{2\text{-}epoxy}$), 2.71 (dd, J = 4.9Hz, J = 2.6Hz,

1 H, CH$_{2\text{-}epoxy}$), 1.95 (s, 3H, CH$_3$), 1.94 (s, 3H, CH$_3$) ppm; ^{13}C NMR (125 MHz, CDCl$_3$): δ =

176.6 (C=O), 66.1 (COO<u>C</u>H$_2$), 55.5 (CBr), 49.2 (CH$_{epoxy}$), 44.6 (CH$_{2\text{-}epoxy}$), 30.8 (CH$_3$) ppm;

FTIR (KBr): υ = 3005 (m), 2932 (w), 1738 (vs), 1464 (m), 1390 (m), 1372 (m), 1344 (w),

1278 (s), 1166 (s), 1109 (s), 1013 (m), 990 (m), 907 (m), 843 (m), 766 (w), 546 (w), 545 (vw)

cm^{-1}; MS (EI, 70eV), m/z (%): 224.0 (11) [M]$^+$, 222.0 (22) [M]$^+$, 211.0 (100) [C$_6$H$_9$BrO$_3$]+,

209.0 (95) [C$_6$H$_9$BrO$_3$]$^+$, 207.0 (29) [C$_7$H$_{10}$BrO$_2$]$^+$, 205.0 (26) [C$_7$H$_{10}$BrO$_2$]$^+$, 151.0 (52)

[C$_4$H$_3$BrO]$^+$, 149.0 (48) [C$_4$H$_3$BrO]$^+$, 143.1 (37) [C$_7$H$_{11}$O$_3$]$^+$, 123.0 (66) [C$_3$H$_6$Br]$^+$, 121.0 (73)

[C$_3$H$_6$Br]$^+$, 69.0 (96) [C$_4$H$_5$O]$^+$; HRMS calcd. for C$_7$H$_{11}$BrO$_3$: 221.9892; found 221.9894.

Analytical data for epoxide **9** matched literature data.[234]

Glycidyl 4-benzoylbenzoate (10)

4-benzoylbenyoic acid (1 g, 4.8 mmol, 1 eq), DIC (1 mL, 6.4 mmol, 1.3 eq), 4-dimethylaminopyridine (DMAP) (40 mg, 0.32 mmol, 0.07 eq) were stirred in 10 mL of dry dichloromethane for 2 h, followed by addition of glycidol (4 mL, 60 mmol, 12.5 eq). The reaction was stirred overnight. Solvent was evaporated *in vacuo*, the residue was dissolved in dichloromethane, washed with half saturated sodium hydrogen carbonate solution and water, and dried over anhydrous magnesium sulfate. After removal of the solvent *in vacuo* the crude product was purified chromatographically on silica gel with hexane/ethyl acetate (1:1 v/v) mixture as an eluent to yield 692 mg (51%) of pure epoxide **10** as a colorless viscous oil. R_f = 0.72. ^{1}H NMR (500 MHz, CDCl$_3$): δ = 8.18 (d, J = 8.4 Hz, 2H, H$_{Ar}$), 7.85 (d, J = 8.4 Hz, 2H, H$_{Ar}$), 7.80 (dd, J = 8.1 Hz, J = 1.2 Hz, 2H, H$_{Ar}$), 7.63 (tt, J = 7.4 Hz, J = 1.2 Hz, 1H, H$_{Ar}$), 7.50 (dd, J = 8.1 Hz, J = 7.4 Hz, 2H, H$_{Ar}$), 4.71 (dd, J = 12.3 Hz, J = 3.0 Hz, 1H, CH$_2$O), 4.21 (dd, J = 12.3 Hz, J = 6.4 Hz, 1H, CH$_2$O), 3.39–3.35 (m, 1H, CH$_{epoxy}$), 2.92 (dd, J = 4.8 Hz, J = 4.4 Hz, 1H, CH$_{2-epoxy}$), 2.75 (dd, J = 4.8 Hz, J = 2.6 Hz, 1H, CH$_{2-epoxy}$) ppm; ^{13}C NMR (125 MHz, CDCl$_3$): δ = 196.1 (C=O), 165.6 (OC=O), 141.7 (C$_{Ar}$), 137.0 (C$_{Ar}$), 133.1 (CH$_{Ar}$), 132.8 (C$_{Ar}$), 130.3 (2C, CH$_{Ar}$), 129.9 (2C, CH$_{Ar}$), 129.8 (2C, CH$_{Ar}$), 128.6 (2C, CH$_{Ar}$), 66.0 (CH$_2$O), 49.5 (CH$_{epoxy}$), 44.8 (CH$_{2-epoxy}$) ppm; FTIR (KBr): υ = 3060 (w), 1724 (s), 1661 (s), 1597 (m), 1578 (w), 1502 (w), 1448 (m), 1405 (m), 1345 (m), 1272 (s), 1181 (w), 1105 (s), 1018 (m), 986 (w), 927 (m), 869 (m), 767 (m), 714 (s), 698 (m), 657 (m) cm^{-1}; MS (EI, 70eV), m/z (%): 282.1 (90) [M]$^+$, 209.1 (100) [C$_{14}$H$_9$O$_2$]$^+$, 205.1 (39) [C$_{11}$H$_9$O$_4$]$^+$, 105.1 (89) [C$_7$H$_5$O]$^+$, 77.1 (33) [C$_6$H$_5$]$^+$; HRMS calcd. for C$_{17}$H$_{14}$O$_4$: 282.0892; found 282.0891.

Glycidaldehyde diethyl acetal (GADEA, 11)

The synthetic procedure was adapted from that which was previously reported.[159] Potassium bicarbonate (2.92 g, 29.2 mmol, 0.15 eq) was suspended in 100 mL of methanol followed by addition of acrolein diethyl acetal (29.3 mL, 192 mmol, 1 eq), benzonitrile (18.2 mL) and hydrogen peroxide (35%, 19.6 mL, 202 mmol, 1.05 eq). The reaction was stirred at 40°C for 8 h. Next, 6.5 mL of hydrogen peroxide was added and the reaction was stirred overnight at room temperature, followed by addition of a new portion of hydrogen peroxide (6.5 mL) and stirring for 20 h. The reaction was quenched by addition of water (130 mL), product was extracted by dichloromethane and the combined organic phase was dried over anhydrous magnesium sulfate. Solvent was evaporated *in vacuo*. n-Hexane was added to the residue and the obtained precipitate was filtered off. After removal of solvent, the crude product was purified by distillation to yield 19.5 g (70%) of epoxide **11** (inseparable mixture of diastereomers) as a colorless liquid. Bp 74–75°C/28 mbar. ^{1}H NMR (500 MHz, CDCl$_3$): δ = 4.32 (d, J = 4.3 Hz, 1H, C$\underline{H}$(OEt)$_2$), 3.77–3.67 (m, 2H, C$\underline{H}_2$CH$_3$), 3.64–3.52 (m, 2H, C$\underline{H}_2$CH$_3$), 3.10–3.06 (m, 1H, CH$_{epoxy}$), 2.79–2.73 (m, 2H, CH$_{2\text{-epoxy}}$), 1.23 (t, J = 7.1 Hz, 3H, CH$_3$), 1.20 (t, J = 7.1 Hz, 3H, CH$_3$) ppm; ^{13}C NMR (125 MHz, CDCl$_3$): δ = 101.4 ($\underline{C}$H(OEt)$_2$), 62.7 ($\underline{C}$H$_2$CH$_3$), 62.2 ($\underline{C}$H$_2$CH$_3$), 51.7 (CH$_{epoxy}$), 43.6 (CH$_{2\text{-epoxy}}$), 15.1 (CH$_3$) ppm. Analytical data for epoxide **11** matched literature data.[159]

Glycidyl propiolate (27)

Propiolic acid (61 μL, 1 mmol, 1 eq) and BtOH·H$_2$O (153 mg, 1 mmol, 1 eq) were dissolved in 9 mL of dichloromethane and cooled to 0°C. Next, DIC (170 μL, 1.1 mmol, 1.1 eq) was added dropwise and the reaction was stirred at 0°C for 4 h. The precipitate was filtered off and glycidol (331 μL, 5 mmol, 5 eq) was added dropwise to the filtrate at 0°C, after which the reaction was stirred at room temperature overnight. The reaction was quenched with water and product was extracted with dichloromethane. Organic extracts were separated, dried over anhydrous magnesium sulfate and solvent was evaporated *in vacuo*. The crude product was purified chromatographically on silica gel with dichloromethane/ethyl acetate (9:1 v/v) mixture as eluent to yield 77 mg (60%) of epoxide **27** as a slightly yellow liquid. R_f = 0.93. ^{1}H NMR (500 MHz, CDCl$_3$): δ = 4.48 (dd, J = 12.2 Hz, J = 3.2 Hz, 1 H, CH$_2$OC=O), 4.06 (dd, J = 12.2 Hz, J = 6.2 Hz, 1 H, CH$_2$OC=O), 3.26–3.21 (m, 1H, CH$_{epoxy}$), 2.94 (s, 1H, CH≡C), 2.86 (dd, J = 4.8 Hz, J = 4.2 Hz, 1 H, CH$_{2\text{-epoxy}}$), 2.67 (dd, J = 4.8 Hz, J = 2.6 Hz, 1 H, CH$_{2\text{-epoxy}}$) ppm. ^{13}C NMR (125 MHz, CDCl$_3$): δ = 152.4 (C=O), 75.8 (<u>C</u>H≡C), 74.2 (CH≡<u>C</u>), 66.6 (<u>C</u>H$_2$OC=O), 48.8 (CH$_{epoxy}$), 44.8 (CH$_{2\text{-epoxy}}$) ppm; FTIR (KBr): υ =3262 (w), 3009 (w), 2121 (m), 1718 (m), 1483 (vw), 1450 (w), 1344 (w), 1228 (m), 1163 (w), 1137 (w), 1074 (vw), 987 (w), 908 (w), 858 (w), 754 (w), 694 (w), 583 (vw) cm^{-1}.

2-Hydroxy-5-methoxy-α,α-diphenyl benzenemethanol (30)

The synthetic procedure was adapted from that which was previously reported.[196] Solution of 26.3 mL (0.25 mol, 4.4 eq) of bromobenzene in 65 mL of dry THF was added dropwise to 5.5 g (0.23 mol, 4 eq) of magnesium in 25 mL of dry THF followed by refluxing for 1 h under an argon atmosphere. After the reaction was cooled to room temperature methyl 2-hydroxy-5-methoxybenzoate (**29**, 10.4 g, 57 mmol, 1 eq) in additional 90 mL of dry THF was added dropwise thereto and the reaction was stirred overnight at room temperature. The reaction was quenched with a saturated solution of ammonium chloride and extracted with diethyl ether. The joint extracts were washed with brine, dried over anhydrous sodium sulfate, filtered and concentrated *in vacuo*. The crude product was purified chromatographically on silica gel with hexane/ethyl acetate (2:1 v/v) mixture as eluent to yield 13.25 g (77%) of the title compound as a white solid. R_f = 0.62. Mp 140–144°C (decomposition). ^{1}H NMR (500 MHz, CDCl$_3$): δ = 7.55 (br s, 1H, OH), 7.37–7.30 (m, 6H, H$_{Ar}$), 7.24–7.20 (m, 4H, H$_{Ar}$), 6.80 (d, J = 8.8 Hz, 1H, H$_{Ar}$), 6.77 (dd, J = 8.8 Hz, J = 3.0 Hz, 1H, H$_{Ar}$), 6.11 (d, J = 3.0 Hz, 1H, H$_{Ar}$), 3.71 (br s, 1H, OH), 3.61 (s, 3H, CH$_3$O) ppm; ^{13}C NMR (125 MHz, CDCl$_3$): δ = 152.3 (C$_{Ar}$), 150.0 (C$_{Ar}$), 144.8 (2C, C$_{Ar}$), 131.2 (C$_{Ar}$), 128.4 (4C, CH$_{Ar}$), 128.1 (2C, CH$_{Ar}$), 127.9 (4C, CH$_{Ar}$), 118.1 (CH$_{Ar}$), 116.7 (CH$_{Ar}$), 113.9 (CH$_{Ar}$), 84.3 (C), 55.7 (CH$_3$) ppm; FTIR (Platinum ATR): υ = 3358 (w), 3172 (w), 1599 (vw), 1490 (m), 1465 (w), 1447 (w), 1381 (w), 1288 (w), 1274 (w),

1212 (m), 1172 (w), 1140 (w), 1079 (vw), 1035 (w), 1010 (m), 936 (w), 905 (vw), 860 (w), 819 (w), 770 (m), 754 (m), 696 (m), 655 (w), 645 (w), 604 (w), 564 (w), 537 (w) cm^{-1}; MS (EI, 70eV), m/z (%): 306.2 (4) [M]$^+$; 289.1 (21) [C$_{20}$H$_{17}$O$_2$]$^+$; 288.1 (90) [C$_{20}$H$_{16}$O$_2$]$^+$; 287.1 (100) [C$_{20}$H$_{15}$O$_2$]$^+$; 211.1 (29) [C$_{14}$H$_{11}$O$_2$]$^+$; HRMS calcd. for C$_{20}$H$_{18}$O$_3$: 306.1256; found 306.1259; elemental analysis calcd (%) for C$_{20}$H$_{18}$O$_3$: C 78.41, H 5.92; found: C 78.35, H 5.95. Analytical data for diol **30** matched literature data.[196]

6-Methoxy-2-vinyl-4,4-diphenyl-4*H*-1,3-benzodioxin (31)

The synthetic procedure was adapted from that which was previously reported for the synthesis of analogous acetals.[196] Diol **30** (4.5 g, 14.7 mmol, 1.5 eq), acrolein (0.65 mL, 9.8 mmol, 1 eq) and *p*-toluenesulfonic acid were stirred in benzene (48 mL) for 3 days at room temperature. The solvent was evaporated *in vacuo* and the crude product was purified chromatographically on silica gel with hexane/ethyl acetate (4:1 v/v) mixture as eluent to yield 3.2 g (95%) of the title compound as a white solid. R$_f$ = 0.68. Mp 105–107°C. ^{1}H NMR (500 MHz, CDCl$_3$): δ = 7.44–7.33 (m, 5H, H$_{Ar}$), 7.29–7.20 (m, 5H, H$_{Ar}$), 6.90 (d, J = 8.9 Hz, 1H, H$_{Ar}$), 6.77 (dd, J = 8.9 Hz, J = 3.0 Hz, 1H, H$_{Ar}$), 6.39 (d, J = 3.0 Hz, 1H, H$_{Ar}$), 6.06 (ddd, J = 17.4 Hz, J = 10.6 Hz, J = 4.7 Hz, 1H, CH$_2$=C$\underline{H}$), 5.54 (ddd, J = 17.4 Hz, J = 1.2 Hz, J = 0.9 Hz, 1H, CH(O)$_2$), 5.41–5.36 (m, 2H, C$\underline{H}_2$=CH), 3.64 (s, 3H, CH$_3$O) ppm; ^{13}C NMR (125 MHz, CDCl$_3$): δ = 153.1 (C$_{Ar}$), 146.3 (C$_{Ar}$), 145.8 (C$_{Ar}$), 144.0 (C$_{Ar}$), 133.9 (CH$_2$=$\underline{C}$H), 129.3 (2C, CH$_{Ar}$), 128.3 (4C, CH$_{Ar}$), 128.2 (CH$_{Ar}$), 128.0 (2C, CH$_{Ar}$), 127.7 (CH$_{Ar}$), 126.1 (C$_{Ar}$), 119.6 ($\underline{C}$H$_2$=CH), 117.9 (CH$_{Ar}$), 115.1 (CH$_{Ar}$), 114.3 (CH$_{Ar}$), 94.0 (CH(O)$_2$), 84.7 (C), 55.7 (CH$_3$O) ppm; FTIR (Platinum ATR): υ = 2995 (vw), 2952 (vw), 2864 (vw), 1613 (vw), 1489 (w), 1461 (vw), 1444 (w), 1413 (w), 1361 (w), 1328 (vw), 1269 (w), 1223 (w), 1200 (w), 1177 (w), 1130 (w), 1066 (w), 1037 (w), 994 (w), 951 (w), 850 (w), 816 (w), 802 (w), 771 (w), 760 (w), 727 (w), 700 (w), 646 (w), 609 (vw), 585 (vw) cm^{-1}; MS (EI, 70eV), m/z (%): 344.2 (2) [M]$^+$; 289.1 (40) [C$_{20}$H$_{17}$O$_2$]$^+$; 288.1 (91) [C$_{20}$H$_{16}$O$_2$]$^+$; 287.1 (100) [C$_{20}$H$_{15}$O$_2$]$^+$; 211.1 (20) [C$_{14}$H$_{11}$O$_2$]$^+$; HRMS calcd. for C$_{23}$H$_{20}$O$_3$: 344.1412; found 344.1415; elemental analysis calcd (%) for C$_{23}$H$_{20}$O$_3$: C 80.21, H 5.85; found: C 80.27, H 5.87.

6-Methoxy-2-oxiranyl-4,4-diphenyl-4H-1,3-benzodioxin (32)

The synthesis was performed according to procedure for the synthesis of epoxide **11** by epoxidation of acetal **31** (1.3 g, 3.7 mmol, 1 eq). The reaction temperature was increased to 60°C. The crude product was purified chromatographically on silica gel with hexane/ethyl acetate (4:1 v/v) mixture as eluent to yield 1.0 g (74%) of epoxide **32** (mixture of diastereomers 1:1.35) as a white solid. R$_f$ = 0.53. ^{1}H NMR (500 MHz, CDCl$_3$): δ = 7.40–7.32 (m, 5H, H$_{Ar}$), 7.30–7.22 (m, 5H, H$_{Ar}$), 6.93 and 6.88 (d, J = 8.9 Hz, 1H, H$_{Ar}$), 6.80–6.75 (m, 1H, H$_{Ar}$), 6.37 and 6.35 (d, J = 3.0 Hz, 1H, H$_{Ar}$), 4.84 and 4.75 (d, J = 4.3 and 4.7 Hz, 1H, CH(O)$_2$), 3.64 (s, 3H, CH$_3$O), 3.38–3.32 (m, 1H, CH$_{epoxy}$), 2.89–2.64 (m, 2H, CH$_{2\text{-epoxy}}$) ppm, ^{13}C NMR (125 MHz, CDCl$_3$): δ = 153.3 (C$_{Ar}$), 145.7 and 145.6 (C$_{Ar}$), 145.4 and 145.3 (C$_{Ar}$), 143.5 and 143.4 (C$_{Ar}$), 129.3 and 129.2 (2C, CH$_{Ar}$), 128.5 and 128.4 (CH$_{Ar}$), 128.4 and 128.4 (2C, CH$_{Ar}$), 128.3 (2C, CH$_{Ar}$), 128.1 and 128.1 (2C, CH$_{Ar}$), 127.9 and 127.9 (CH$_{Ar}$), 126.1 and 126.0 (C$_{Ar}$), 118.0 and 117.8 (CH$_{Ar}$), 115.1 and 115.0 (CH$_{Ar}$), 114.4 (CH$_{Ar}$), 94.7 and 94.2 (CH(O)$_2$), 84.8 and 84.6 (C), 55.7 (CH$_3$O), 51.8 and 51.7 (CH$_{epoxy}$), 43.8 and 43.8 (CH$_{2\text{-epoxy}}$) ppm; FTIR (Platinum ATR): υ = 3059 (vw), 2995 (vw), 2915 (vw), 2835 (vw), 1615 (vw), 1491 (m), 1466 (w), 1445 (m), 1415 (w), 1312 (w), 1270 (w), 1249 (w), 1221 (m), 1182 (w), 1143 (w), 1074 (m), 1033 (m), 989 (m), 939 (w), 926 (m), 850 (w), 831 (w), 810 (m), 782 (w), 768 (m), 756 (w), 727 (w), 697 (m), 632 (w) cm^{-1}; MS (EI, 70eV), m/z (%): 360.1 (4) [M]$^+$; 289.1 (13) [C$_{20}$H$_{17}$O$_2$]$^+$; 288.1 (63) [C$_{20}$H$_{16}$O$_2$]$^+$; 287.1 (58) [C$_{20}$H$_{15}$O$_2$]$^+$; 211.1 (12) [C$_{14}$H$_{11}$O$_2$]$^+$; 58.0 (34) [C$_3$H$_6$O]$^+$; 43.0 (100) [C$_2$H$_3$O]$^+$; HRMS calcd. for C$_{23}$H$_{20}$O$_4$: 360.1362; found 360.1360.

Allyl 2-Nitrobenzyl Ether (48)

The synthetic procedure was adapted from that which was previously reported.[235] 1.152 g (95% suspension in mineral oil, 45.6 mmol, 3.3 eq) of NaH was washed from mineral oil with hexane, suspended in 30 mL of hexane under an argon atmosphere and cooled down to 0°C. 15 mL (0.22 mol, 16.0 eq) of allyl alcohol was added dropwise thereto. Next solution of *o*-nitrobenzyl bromide (3.0 g, 13.8 mmol, 1 eq) in 39 mL of allyl alcohol was added dropwise to the reaction at 0°C. The reaction was stirred overnight at room temperature, quenched with 9 mL of 2 M hydrochloric acid and 42 mL of water and extracted with chloroform. The joint extracts were dried over anhydrous magnesium sulfate, filtered and concentrated *in vacuo*. The crude product was purified chromatographically on silica gel with hexane/ethyl acetate (4:1 v/v) mixture as eluent to yield 1.93 g (72%) of compound the title compound as a yellow liquid. $R_f = 0.63$. ^{1}H NMR (500 MHz, CDCl$_3$): δ = 8.06 (dd, J = 8.2 Hz, J = 1.2 Hz, 1H, H$_{Ar}$), 7.83 (dd, J = 7.9 Hz, J = 0.9 Hz, 1H, H$_{Ar}$), 7.65 (td, J = 7.6 Hz, J = 1.2 Hz, 1H, H$_{Ar}$), 7.45–7.41 (m, 1H, H$_{Ar}$), 6.02–5.92 (m, 1H, C$\underline{H}$=CH$_2$), 5.35 (ddt, J = 17.2 Hz, J = 2.8 Hz, J = 1.4 Hz, 1H, CH=C$\underline{H}_2$), 5.24 (ddt, J = 10.4 Hz, J = 2.8 Hz, J = 1.4 Hz, 1H, CH=C$\underline{H}_2$), 4.90 (s, 2H, ArCH$_2$O), 4.13 (dt, J = 5.5 Hz, J = 1.4 Hz, 1H, OC$\underline{H}_2$CH=CH$_2$) ppm; ^{13}C NMR (125 MHz, CDCl$_3$): δ = 147.4 (C$_{Ar}$), 135.4 (C$_{Ar}$), 134.3 (C$\underline{H}$=CH$_2$), 133.8 (CH$_{Ar}$), 128.8 (CH$_{Ar}$), 128.0 (CH$_{Ar}$), 124.8 (CH$_{Ar}$), 117.6 (CH=C$\underline{H}_2$), 72.1 (OCH$_2$CH=CH$_2$), 68.8 (Ar$\underline{C}$H$_2$O) ppm; Analytical data for ether **48** matched literature data.[235]

Glycidyl 2-Nitrobenzyl Ether (49)

The synthetic procedure was adapted from that which was previously reported.[236] Ether **48** (1.5 g, 7.8 mmol, 1 eq) was dissolved in 50 mL of chloroform under an argon atmosphere and cooled down to 0°C. Next, mCPBA (2 g, 11.6 mmol, 1.5 eq) was added thereto and the reaction was stirred for 5 days at room temperature. The reaction was quenched with 3 M NaOH, washed with water, dried over anhydrous magnesium sulfate, filtered and concentrated *in vacuo*. The crude product was purified chromatographically on silica gel with hexane/ethyl acetate (3:1 v/v) mixture as eluent to yield 1.35 g (83%) of epoxide **49** as a colorless liquid. R_f = 0.45. ^{1}H NMR (500 MHz, CDCl$_3$): δ = 8.03 (dd, J = 8.2 Hz, J = 1.2 Hz, 1H, H$_{Ar}$), 7.78 (dd, J = 7.9 Hz, J = 0.9 Hz, 1H, H$_{Ar}$), 7.63 (td, J = 7.6 Hz, J = 1.2 Hz, 1H, H$_{Ar}$), 7.44–7.39 (m, 1H, H$_{Ar}$), 4.93 (s, 2H, ArCH$_2$O), 3.88 (dd, J = 11.4 Hz, J = 2.7 Hz, 1H, OCH$_2$), 3.50 (dd, J = 11.4 Hz, J = 6.0 Hz, 1H, OCH$_2$), 3.23–3.19 (m, 1H, CH$_{epoxy}$), 2.81 (dd, J = 5.0 Hz, J = 4.2 Hz, 1H, CH$_{2\text{-epoxy}}$), 2.64 (dd, J = 5.0 Hz, J = 2.7 Hz, 1H, CH$_{2\text{-epoxy}}$) ppm; ^{13}C NMR (125 MHz, CDCl$_3$): δ = 147.3 (C$_{Ar}$), 134.7 (C$_{Ar}$), 133.8 (CH$_{Ar}$), 128.7 (CH$_{Ar}$), 128.1 (CH$_{Ar}$), 124.7 (CH$_{Ar}$), 71.8 (OCH$_2$), 69.8 (ArCH$_2$O), 50.7 (CH$_{epoxy}$), 44.3 (CH$_{2\text{-epoxy}}$) ppm; Analytical data for ether **49** matched literature data.[236]

6.2.3   Other Small Molecular Weight Compounds

(6-Methoxy-4,4-diphenyl-4H-1,3-benzodioxin-2-yl)-2-(2-methoxyethoxy)ethanol (34)

The synthesis was performed according to procedure B for polymerization (Chapter 6.3.1.2), but with initiator to monomer feed ratio 1:0.9. The crude product was purified chromatographically on silica gel with hexane/ethyl acetate (1:2 v/v) mixture as eluent to yield the title compound (70%) (mixture of diastereomers D1 and D2 in a ratio 1:2.3 correspondingly) as a slightly yellow viscous oil.   The diastereomers were separated chromatographically and analyzed.

D1: R_f = 0.62. ^{1}H NMR (500 MHz, CDCl$_3$): δ = 7.38–7.32 (m, 5H, H$_{Ar}$), 7.29–7.21 (m, 5H, H$_{Ar}$), 6.89 (d, J = 8.9 Hz, 1H, H$_{Ar}$), 6.76 (dd, J = 8.9 Hz, J = 3.0 Hz, 1H, H$_{Ar}$), 6.35 (d, J = 3.0 Hz, 1H, H$_{Ar}$), 5.02 (d, J = 3.8 Hz, 1H, CH(O)$_2$), 4.09–4.04 (m, 1H, C$\underline{H}$OH), 3.77–3.68 (m, 2H, OC$\underline{H}_2$CH), 3.66–3.56 (m, 5H, CH$_3$OAr + 2×OCH$_2$CH$_2$O), 3.52–3.42 (m, 2H, 2×OCH$_2$CH$_2$O), 3.32 (s, 3H, C$\underline{H}_3$OCH$_2$), 2.79 (br s, 1H, OH) ppm; ^{13}C NMR (125 MHz, CDCl$_3$): δ = 153.2 (C$_{Ar}$), 146.0 (C$_{Ar}$), 145.6 (C$_{Ar}$), 143.6 (C$_{Ar}$), 129.4 (2C, CH$_{Ar}$), 128.3 (CH$_{Ar}$), 128.2 (4C, CH$_{Ar}$), 128.1 (2C, CH$_{Ar}$), 127.8 (CH$_{Ar}$), 125.9 (C$_{Ar}$), 118.0 (CH$_{Ar}$), 115.0 (CH$_{Ar}$), 114.3 (CH$_{Ar}$), 93.9 (CH(O)$_2$), 84.6 (C), 72.0 (CH$_2$), 71.6 (CHOH), 71.1 (CH$_2$), 70.9 (CH$_2$), 59.1 ($\underline{C}$H$_3$OCH$_2$), 55.7 ($\underline{C}$H$_3$OAr) ppm.

D2: R_f = 0.55. ^{1}H NMR (500 MHz, CDCl$_3$): δ = 7.38–7.30 (m, 5H, H$_{Ar}$), 7.28–7.21 (m, 5H, H$_{Ar}$), 6.92 (d, J = 8.9 Hz, 1H, H$_{Ar}$), 6.76 (dd, J = 8.9 Hz, J = 3.0 Hz, 1H, H$_{Ar}$), 6.34 (d, J = 3.0 Hz, 1H, H$_{Ar}$), 6.97 (d, J = 5.3 Hz, 1H, CH(O)$_2$), 4.11–4.05 (m, 1H, C$\underline{H}$OH), 3.80 (dd, J = 10.3 Hz, J = 3.3 Hz, 1H, OC$\underline{H}_2$CH), 3.65–3.53 (m, 6H, 1×OC$\underline{H}_2$CH + CH$_3$OAr + 2×OCH$_2$CH$_2$O), 3.46 (t, J = 4.8 Hz, 2H, 2×OCH$_2$CH$_2$O), 3.35 (s, 3H, C$\underline{H}_3$OCH$_2$), 2.14 (br s, 1H, OH) ppm; ^{13}C NMR (125 MHz, CDCl$_3$): δ = 153.2 (C$_{Ar}$), 145.9 (C$_{Ar}$), 145.7 (C$_{Ar}$), 143.6 (C$_{Ar}$), 129.5 (2C, CH$_{Ar}$), 128.4 (CH$_{Ar}$), 128.2 (4C, CH$_{Ar}$), 128.0 (2C, CH$_{Ar}$), 127.8 (CH$_{Ar}$), 125.9 (C$_{Ar}$), 118.0 (CH$_{Ar}$), 115.0 (CH$_{Ar}$), 114.3 (CH$_{Ar}$), 94.2 (CH(O)$_2$), 84.6 (C), 71.9 (CH$_2$), 71.8 (CHOH), 71.0 (CH$_2$), 70.8 (CH$_2$), 59.1 ($\underline{C}$H$_3$OCH$_2$), 55.7 ($\underline{C}$H$_3$OAr) ppm.

FTIR (Platinum ATR): υ = 3405 (vw), 2903 (vw), 1599 (vw), 1492 (w), 1446 (w), 1272 (vw), 1224 (w), 1198 (vw), 1101 (w), 1035 (w), 995 (w), 947 (vw), 871 (vw), 849 (vw), 809 (vw), 758 (w), 700 (w), 647 (vw), 629 (vw), 585 (vw) cm^{-1}, MS (EI, 70eV), m/z (%): 436.1 (3) [M]$^+$, 289.1 (31) [C$_{20}$H$_{17}$O$_2$]$^+$, 288.1 (100) [C$_{20}$H$_{16}$O$_2$]$^+$, 287.1 (72) [C$_{20}$H$_{15}$O$_2$]$^+$, 211.1 (13) [C$_{14}$H$_{11}$O$_2$]$^+$, 43.0 (37) [C$_2$H$_3$O]$^+$; HRMS calcd. for C$_{26}$H$_{28}$O$_6$: 436.1886; found 436.1884.

(4-Nitrophenyl)methyl mercaptoacetate (50)

Thioglycolic acid (8.3 mL, 0.12 mol, 2 eq) and *o*-nitrobenzyl alcohol (9.18 g, 0.06 mol, 1 eq) were dissolved in toluene (90 mL) and conc. sulfuric acid (0.3 mL) was added thereto. The reaction was refluxed for 2 days and solvent was removed *in vacuo*. The crude product was purified chromatographically on silica gel with hexane/ethyl acetate (4:1 v/v) mixture as eluent to yield 8.69 g (64%) of the title compound as a yellow liquid. R$_f$ = 0.32. ^{1}H NMR (500 MHz, CDCl$_3$): δ = 8.09 (dd, J = 8.3 Hz, J = 0.9 Hz, 1H, H$_{Ar}$), 7.69–7.60 (m, 2H, H$_{Ar}$), 7.49 (td, J = 8.3 Hz, J = 1.7 Hz, 1H, H$_{Ar}$), 5.55 (s, 2H, ArCH$_2$O), 3.34 (d, J = 8.4 Hz, 2H, CH$_2$SH), 2.05 (t, J = 8.4 Hz, 1H, CH$_2$SH) ppm; ^{13}C NMR (125 MHz, CDCl$_3$): δ = 170.3 (C=O), 147.5 (C$_{Ar}$), 133.9 (CH$_{Ar}$), 131.6 (C$_{Ar}$), 129.1 (2C, CH$_{Ar}$), 125.2 (CH$_{Ar}$), 63.9 (ArCH$_2$O), 26.4 (CH$_2$SH) ppm; FTIR (Platinum ATR): υ = 3113 (vw), 2957 (vw), 2854 (vw), 2562 (w), 1842 (vw), 1724 (m), 1611 (w), 1575 (w), 1517 (m), 1439 (w), 1414 (w), 1375 (w), 1340 (m), 1309 (w), 1280 (m), 1224 (m), 1142 (m), 1080 (w), 1048 (w), 1023 (m), 979 (w), 965 (w), 936 (w), 860 (w), 795 (m), 730 (m), 671 (m), 600 (w), 587 (w) cm^{-1}; MS (EI, 70eV), m/z (%): 227 (0.05) [M]$^+$, 136 (100) [C$_7$H$_6$NO$_2$]$^+$, 105.0 (21) [C$_3$H$_5$O$_2$S]$^+$, 92.0 (35) [C$_7$H$_8$]$^+$, 78 (67) [C$_6$H$_6$]$^+$, 77 (41) [C$_6$H$_5$]$^+$, 47 (42) [CH$_3$S]$^+$; HRMS calcd. for C$_9$H$_9$NSO$_4$: 227.0252, found 227.0255; elemental analysis calcd (%) for C$_9$H$_9$NSO$_4$: C 47.57, H 3.99, N 6.16, S 14.11; found: C 47.46, H 3.88, N 6.12, S 13.83.

6.3   Synthesis and Characterization of Polymers

6.3.1   General Procedures

6.3.1.1   General Procedure A for Polymerization of EO

Deprotonating agent was added to the solution of an alcohol in dry THF at room temperature or at –78°C. In case of PN the complete formation of an alkoxide was indicated by the maintenance of the solution's green color. EO was introduced into the reaction vessel from a graduated flask ether by distillation to the frozen to –78°C initiating solution or via cannula at 0°C and the reaction was stirred for 72 h at a defined temperature. Polymerization was terminated by the addition of methanol (in case of the presence of acid sensitive functional groups) or acidified methanol, and solvent was removed *in vacuo*. EO conversions were determined gravimetrically. The obtained polymer was further purified by precipitation in cold diethyl ether.

6.3.1.2   General Procedures B for Polymerization of Functional Epoxides

Polymerizations were performed in bulk or in solution. Commercial potassium *tert*-butoxide was used as initiator, or initiating alkoxide was prepared *in situ* from 2-methoxyethanol (pure – bulk polymerization or in dry THF solution – polymerization in solution) and a deprotonating agent, or macroinitiator was prepared *in situ* from mPEG/PEG and PN as a deprotonating agent. In the case of PN, the complete formation of 2-methoxyethanolate was indicated by the maintenance of the solution's green color. In the case of bulk polymerization the obtained alkoxide was dried *in vacuo*. Next, the monomer(s) was(were) introduced into the reaction vessel with a gastight syringe and the reaction was stirred for a given period of time at a defined temperature. In case of polymerization in solution the monomer concentration in THF was approximately 25 w/v%. Polymerization was terminated by the addition of methanol (if acid sensitive functional groups were present) or acidified methanol, and solvent was removed *in vacuo*. Monomer conversion was calculated based on ^{1}H NMR spectra of crude polymer. The obtained polymer was further purified chromatographically on a silica gel column with gradient elution (chloroform to chloroform/methanol mixture as an eluent).

6.3.1.3   General Procedure C for Cleavage of Protecting Groups in Multifunctional PEGs [99]

Polymer (1 eq of acetal protected group) was dissolved in THF (0.8 w/v%) and stirred with hydrochloric acid (37%, 10 eq) for 10 min. Polymer precipitated as a colorless oil. Solvent was decanted, the residue was washed with THF three times and dried *in vacuo*.

6.3.1.4   General Procedure D for Thiol-Ene Post-polymerization Modification[125]

Thiol (5 eq) and AIBN (0.15 eq) were added to the solution of purified AGE-(co)polymer (1 eq of C=C) in dry THF under an argon atmosphere. The mixture was refluxed for 5 h. Completion of the reaction was confirmed by ^{1}H NMR spectrometry by disappearance of allyl group signal. The solvent was removed *in vacuo* and the residue was purified on silica gel column with gradient elution (chloroform to chloroform/methanol mixture as an eluent).

6.3.1.5   General Procedure for Polyhydrazide Syntheses[125]

Polyester (1 eq of ester groups) was refluxed with 80% water solution of hydrazine hydrate (100 eq) in THF for 5 h. The solvent was evaporated and the product was purified as specified for each polymer.

6.3.2   α,ω-Heterotelechelic PEGs

Methoxy poly(ethylene glycol) (mPEG, 1)

$$MeO \left[\!\!\! \diagdown \!\!\! \diagup \!\!\! O \right]_n H$$

The synthesis was performed according to procedure A (Chapter 6.3.1.1).

White solid. ^{1}H NMR (500 MHz, CDCl$_3$): δ = 3.80–3.47 (m, CH$_{2\text{-PEG}}$), 3.37 (s, 3H, CH$_3$O) ppm; ^{13}C NMR (125 MHz, CDCl$_3$): δ = 70.7 (CH$_{2\text{-PEG}}$), 61.4 (CH$_2$OH), 59.2 (CH$_3$O) ppm.

4-(Dimethoxymethyl)benzyl poly(ethylene glycol) (3d)

The synthesis was performed according to procedure A (Chapter 6.3.1.1).

White solid. ^{1}H NMR (500 MHz, CDCl$_3$): δ = 7.41 (d, 2H, H$_{Ar}$), 7.33 (d, 2H, H$_{Ar}$), 5.37 (s, 1H, C$\underline{H}$(OMe)$_2$), 4.56 (s, 2H, ArC$\underline{H}_2$O), 3.79–3.47 (m, CH$_{2\text{-PEG}}$), 3.31 (s, 6H, CH$_3$O) ppm; ^{13}C NMR (125 MHz, CDCl$_3$): δ = 138. 7 (C$_{Ar}$), 137.5 (C$_{Ar}$), 127.7 (2C, CH$_{Ar}$), 126.9 (2C, CH$_{Ar}$), 103.2 (C$\underline{H}$(OMe)$_2$), 73.1 (ArC$\underline{H}_2$O), 61.4 (CH$_2$OH), 52.8 (CH$_3$O) ppm.

6.3.3 Homomultifunctional PEGs

***tert*-Butoxy poly(allyl glycidyl ether) (PAGE, 12a)**

The synthesis was performed according to procedure B (Chapter 6.3.1.2).

Slightly yellow oil. ^{1}H NMR (500 MHz, CDCl$_3$): δ = 5.93–5.82 (m, 1H, C$\underline{H}$=CH$_2$), 5.25 (d, 1H, CH=C$\underline{H}_2$), 5.14 (d, 1H, CH=C$\underline{H}_2$), 3.98 (d, 2H, OC$\underline{H}_2$ CH=CH$_2$), 3.67–3.43 (m, 5H, CH$_{2\text{-PEG}}$ + CH$_{PEG}$ + CH$_{PEG}$C$\underline{H}_2$O), 1.16 (s, 9H, CH$_3$) ppm.

2-Methoxyethoxy poly(allyl glycidyl ether) (PAGE, 12b)

The synthesis was performed according to procedure B (Chapter 6.3.1.2).

Slightly yellow oil. R$_f$ = 0.68 (chloroform/methanol, 10:0.7 v/v). ^{1}H NMR (500 MHz, CDCl$_3$): δ = 5.93–5.82 (m, 1H, C$\underline{H}$=CH$_2$), 5.25 (d, 1H, CH=C$\underline{H}_2$), 5.14 (d, 1H, CH=C$\underline{H}_2$),

3.98 (d, 2H, OC$\underline{H}_2$CH=CH$_2$), 3.67–3.43 (m, 5H, CH$_{2\text{-PEG}}$ + CH$_{\text{PEG}}$ + CH$_{\text{PEG}}$C$\underline{H}_2$O), 3.36 (s, 3H, CH$_3$O) ppm; ^{13}C NMR (125 MHz, CDCl$_3$): δ = 135.1 (CH$_2$=$\underline{C}$H), 116.9 ($\underline{C}$H$_2$=CH), 79.0–78.8 (CH$_{\text{PEG}}$), 72.4 (O$\underline{C}$H$_2$CH=CH$_2$), 70.4–69.9 (2C, CH$_{2\text{-PEG}}$ + CH$_{\text{PEG}}$C$\underline{H}_2$O), 59.2 (CH$_3$O). Analytical data for polymer **12b** matched literature data.[97]

2-Methoxyethoxy poly(benzyl glycidyl ether) (PbnGE, 14)

The synthesis was performed according to procedure B (Chapter 6.3.1.2).

Colorless oil. R$_f$ = 0.86 (chloroform/methanol, 10:0.7 v/v). ^{1}H NMR (500 MHz, CDCl$_3$): δ = 7.60–7.15 (m, 5H, H$_{\text{Ar}}$), 4.43 (s, 2H, ArC$\underline{H}_2$O), 3.75–3.40 (m, 5H, CH$_{2\text{-PEG}}$ + CH$_{\text{PEG}}$ + CH$_{\text{PEG}}$C$\underline{H}_2$O), 3.32 (s, 3H, CH$_3$O) ppm; ^{13}C NMR (125 MHz, CDCl$_3$): δ = 138.6 (C$_{\text{Ar}}$), 128.4 (2C, CH$_{\text{Ar}}$), 127.7 (2C, CH$_{\text{Ar}}$), 127.6 (CH$_{\text{Ar}}$), 79.0–78.7 (CH$_{\text{PEG}}$), 73.4 (Ar$\underline{C}$H$_2$O), 70.6–69.7 (2C, CH$_{2\text{-PEG}}$ + CH$_{\text{PEG}}$C$\underline{H}_2$O) ppm. Analytical data for polymer **14** matched literature data.[115]

2-Methoxyethoxy poly(ethoxyethyl glycidyl ether) (PEEGE, 15)

The synthesis was performed according to procedure B (Chapter 6.3.1.2).

Colorless oil. R$_f$ = 0.47 (chloroform/methanol, 10:0.7 v/v). ^{1}H NMR (500 MHz, CDCl$_3$): δ = 4.73–4.66 (q, 1H, C$\underline{H}$CH$_3$), 3.76–3.40 (m, 7H, CH$_{2\text{-PEG}}$ + CH$_{\text{PEG}}$ + CH$_{\text{PEG}}$C$\underline{H}_2$O + C$\underline{H}_2$CH$_3$), 3.36 (s, 3H, CH$_3$O), 1.28 (d, 3H, CHC$\underline{H}_3$), 1.18 (t, 3H, CH$_2$C$\underline{H}_3$) ppm, ^{13}C NMR (125 MHz, CDCl$_3$): δ = 100.0 and 99.8 (CH$_3$$\underline{C}$H), 79.2–78.7 (CH$_{\text{PEG}}$), 70.4–69.8 (CH$_{2\text{-PEG}}$), 65.3–64.4 (CH$_{\text{PEG}}C\underline{H}_2$O), 60.9 ($\underline{C}H_2CH_3$), 59.2 (CH$_3$O), 19.9 ($\underline{C}H_3$CH), 15.5 (CH$_2$$\underline{C}H_3$) ppm. Analytical data for polymer **15** matched literature data.[99]

2-Methoxyethoxy poly(glycidaldehyde diethyl acetal) (PGADEA, 16)

The synthesis was performed according to procedure B (Chapter 6.3.1.2).

Slightly yellow oil. R_f = 0.42 (chloroform/methanol, 10:0.7 v/v). ^{1}H NMR (500 MHz, CDCl$_3$): δ = 4.53–4.39 (m, 1H, C$\underline{H}$(OEt)$_2$), 3.93–3.37 (m, 7H, CH$_{2\text{-PEG}}$ + CH$_{\text{PEG}}$ + 2 × C$\underline{H}_2$CH$_3$), 3.35 (s, 3H, CH$_3$O), 1.25–1.14 (CH$_2$C$\underline{H}_3$) ppm; ^{13}C NMR (125 MHz, CDCl$_3$): δ = 103.0–102.5 ($\underline{C}$H(OEt)$_2$), 81.8–80.9 (CH$_{\text{PEG}}$), 71.6–70.4 (CH$_{2\text{-PEG}}$), 63.9–63.2 (2C, $\underline{C}$H$_2$CH$_3$), 59.1 (CH$_3$O), 15.7–15.4 (CH$_2\underline{C}$H$_3$) ppm.

Polyglycidol (PG)

The synthesis was performed according to procedure C (Chapter 6.3.1.3) from polymer **15** (M$_n$ 6100 g/mol, 0.3 g, 2.0 mmol of EEGE) to yield 114 mg (75%) of the title polymer as a colorless oil. ^{1}H NMR (500 MHz, DMSO-d_3): δ = 4.30 (br s, 1H, OH), 3.62–3.33 (m, 5H, CH$_{2\text{-PEG}}$ + CH$_{\text{PEG}}$ + CH$_{\text{PEG}}$C$\underline{H}_2$O), 3.24 (s, 3H, CH$_3$O) ppm; ^{13}C NMR (125 MHz, DMSO-d_3): δ = 8.2–79.9 (CH$_{\text{PEG}}$), 69.6–69.1 (CH$_{2\text{-PEG}}$), 60.9 (CH$_2$OH), 58.2 (CH$_{\text{PEG}}\underline{C}H_2$O). Analytical data for PG matched literature data.[99]

Photosensitive protected aldehyde poly(ethylene glycol) (PPAPEG, 33)

The synthesis was performed according to procedure B (Chapter 6.3.1.2).

. R_f = 0.98 (chloroform/methanol, 10:0.7 v/v). ^{1}H NMR (500 MHz, CDCl$_3$): δ = 7.40–6.75 (10H, H$_{Ar}$), 6.75–6.36 (2H, H$_{Ar}$), 6.28 (1H, H$_{Ar}$), 4.91 (1H, CH(O)$_2$), 4.25–3.20 (6H, CH$_{2\text{-PEG}}$ + CH$_{PEG}$ + CH$_3$OAr), 3.16 (CH$_3$O) ppm; ^{13}C NMR (125 MHz, CDCl$_3$): δ = 153.0–152.6 (C$_{Ar}$), 146.5–145.5 (2C, C$_{Ar}$), 143.8–143.2 (C$_{Ar}$), 129.5–129.1 (2C, CH$_{Ar}$), 128.1 (5C, CH$_{Ar}$), 127.9 (2C, CH$_{Ar}$), 127.6–127.2 (CH$_{Ar}$), 126.2–125.8 (C$_{Ar}$), 118.3–118.0 (CH$_{Ar}$), 115.1–114.7 (CH$_{Ar}$), 114.2–113.7 (CH$_{Ar}$), 94.3–93.8 (CH(O)$_2$), 84.5–84.1 (C), 80.9–79.7 (CH$_{PEG}$), 70.9–70.1 (CH$_{2\text{-PEG}}$), 55.6 (CH$_3$OAr) ppm; FTIR (spin coated film on a gold wafer): υ = 3252, 3060, 3032, 2932, 1617, 1600, 1496, 1465, 1447, 1274, 1230, 1144, 1124, 1077, 1046, 998 cm^{-1}.

Ester poly(ethylene glycol) (40)

The synthesis was performed according to procedure D (Chapter 6.3.1.4) from polymer **12b** (M$_n$ 11500 g/mol, 0.2 g, 1.75 mmol of allyl groups) resulting in 99.9% conversion of allyl groups into ester groups. The crude polymer was purified chromatographically on silica gel to yield 0.29 g (76%) of the title polymer as a colorless viscous oil. R_f = 0.70 (chloroform/methanol, 10:0.7 v/v). GPC (THF): M$_n$ 15100 g/mol, M$_w$ 16000 g/mol, PDI 1.07. ^{1}H NMR (500 MHz, CDCl$_3$): δ = 3.73 (s, 3H, CH$_3$OC=O), 3.64–3.40 (m, 7H, CH$_{2\text{-PEG}}$ + CH$_{PEG}$ + CH$_{PEG}$C$\underline{H}_2$O + OC$\underline{H}_2$CH$_2$), 3.37 (s, 3H, CH$_3$O), 3.23 (s, 2H, SCH$_2$C=O), 2.70 (t, 2H, SC$\underline{H}_2$CH$_2$), 1.85 (m, 2H, CH$_2$C$\underline{H}_2$CH$_2$) ppm; ^{13}C NMR (125 MHz, CDCl$_3$): δ = 170.9 (C=O),

79.0–78.8 (CH$_{PEG}$), 71.1–70.0 (2C, CH$_{2-PEG}$ + CH$_{PEG}$CH$_2$O), 69.6 (OCH$_2$CH$_2$), 52.4 (CH$_3$OC=O), 33.4 (SCH$_2$C=O), 29.5 (SCH$_2$CH$_2$), 29.2 (CH$_2$CH$_2$CH$_2$) ppm.

Hydrazide poly(ethylene glycol) (PHZ, 41)

The synthesis was performed according to procedure E (Chapter 6.3.1.5) from polymer **40** (M$_n$ 15100 g/mol, 0.1 g, 0.45 mmol of ester groups) resulting in 99.9% conversion of ester groups into hydrazide groups. The crude product was dialyzed against water for 48 h and dried *in vacuo* to yield 87 mg (87%) of the title polymer as a transparent gummy oil.

For comparison of PHZ and P(HZ-*co*-BnGE) (**47**) solubility lower molecular weight PHZ with was synthesized from polymer **40** (M$_n$ 9400 g/mol).

^{1}H NMR (500 MHz, DMSO-d_6): δ = 9.11 (s, 1H, NH$_2$NH), 4.29 (broad, s, 2H, NH$_2$NH), 3.65–3.20 (m, 7H, CH$_{2-PEG}$ + CH$_{PEG}$ + CH$_{PEG}$CH$_2$O + OCH$_2$CH$_2$), 3.25 (s, 3H, CH$_3$O), 3.04 (s, 2H, SCH$_2$C=O), 2.60 (t, 2H, SCH$_2$CH$_2$), 1.76 (m, 2H, CH$_2$CH$_2$CH$_2$) ppm; ^{13}C NMR (125 MHz, DMSO-d_6): δ = 168.5 (C=O), 78.3–78.0 (CH$_{PEG}$), 70.4–69.3 (CH$_{2-PEG}$), 69.5–69.2 (CH$_{PEG}$CH$_2$O), 69.1 (OCH$_2$CH$_2$), 32.6 (SCH$_2$C=O), 28.9 (SCH$_2$CH$_2$), 28.6 (CH$_2$CH$_2$CH$_2$) ppm.

o-Nitrobenzyl poly(ethylene glycol) (NBPEG, 51)

The synthesis was performed according to procedure D (Chapter 6.3.1.4) from polymer **12b** (M$_n$ 11600 g/mol, 0.2 g, 1.8 mmol of allyl groups) resulting in 99.9% conversion of allyl groups into NB groups. The crude polymer was purified chromatographically on silica gel to yield 0.48 g (80%) of the title polymer as a yellow viscous oil. R$_f$ = 0.65 (chloroform/methanol, 10:0.7 v/v). GPC (DMAc): M$_n$ 22100 g/mol, M$_w$ 27500 g/mol, PDI

1.24. ^{1}H NMR (500 MHz, CDCl$_3$): δ = 8.07 (d, 1H, H$_{Ar}$), 7.68–7.62 (m, 2H, H$_{Ar}$), 7.52–7.45 (m, 1H, H$_{Ar}$), 5.52 (s, 2H, ArCH$_2$O), 3.70–3.35 (m, 7H, CH$_{2\text{-PEG}}$ + CH$_{PEG}$ + CH$_{PEG}$C$\underline{H}_2$O + OC$\underline{H}_2$CH$_2$), 3.28 (s, 2H, SCH$_2$C=O), 2.68 (t, 2H, SC$\underline{H}_2$CH$_2$), 1.82 (m, 2H, CH$_2$C$\underline{H}_2$CH$_2$) ppm; ^{13}C NMR (125 MHz, CDCl$_3$): δ = 169.9 (C=O), 147.6 (C$_{Ar}$), 134.1 (CH$_{Ar}$), 132.0 (C$_{Ar}$), 129.2 (CH$_{Ar}$), 129.1 (CH$_{Ar}$), 125.2 (CH$_{Ar}$), 79.2–78.4 (CH$_{PEG}$), 71.4–70.8 (CH$_{2\text{-PEG}}$), 70.3–70.0 (CH$_{PEG}$C$\underline{H}_2$O), 69.7 (OC$\underline{H}_2$CH$_2$), 63.6 (ArC$\underline{H}_2$O), 33.4 (SC$\underline{H}_2$C=O), 29.5 (SC$\underline{H}_2$CH$_2$), 29.1 (CH$_2$C$\underline{H}_2$CH$_2$) ppm; FTIR (spin coated film on a gold wafer): υ = 2922, 2868, 1741, 1614, 1579, 1530, 1448, 1373, 1346, 1272, 1132, 1007, 860 cm^{-1}.

Oxidized o-Nitrobenzyl poly(ethylene glycol) (ONBPEG, 52)

Polymer **51** (M$_n$ 22100 g/mol, 150 mg, 0.44 mmol of thioether groups, 1 eq) was dissolved in 6 mL of acetic acid and 2 mL of acetonitrile and cooled down to 0°C. Next, hydrogen peroxide solution (30%, 0.4 mL, 3.9 mmol, 9 eq) was added thereto dropwise and the reaction was stirred at 0°C for 6 h. The reaction was slowly poured into 5% NaHCO$_3$ solution, extracted with dichloromethane, dried over anhydrous sodium sulfate, filtered and concentrated *in vacuo* to yield 147 mg (94%) of the title polymer as a yellow viscous oil. GPC (DMAc): M$_n$ 26100 g/mol, M$_w$ 34200 g/mol, PDI 1.31. ^{1}H NMR (500 MHz, CDCl$_3$): δ = 8.07 (d, 1H, H$_{Ar}$), 7.71–7.62 (m, 2H, H$_{Ar}$), 7.47 (m, 1H, H$_{Ar}$), 5.57 (s, 2H, ArCH$_2$O), 3.81 (dd, 2H, SCH$_2$C=O), 3.70–3.40 (m, 7H, CH$_{2\text{-PEG}}$ + CH$_{PEG}$ + CH$_{PEG}$C$\underline{H}_2$O + OC$\underline{H}_2$CH$_2$), 2.96 (m, 2H, SC$\underline{H}_2$CH$_2$), 2.0 (m, 2H, CH$_2$C$\underline{H}_2$CH$_2$) ppm; ^{13}C NMR (125 MHz, CDCl$_3$): δ = 165.2 (C=O), 147.4 (C$_{Ar}$), 134.3 (CH$_{Ar}$), 131.2 (C$_{Ar}$), 129.6 (CH$_{Ar}$), 129.3 (CH$_{Ar}$), 125.2 (CH$_{Ar}$), 79.0–78.5 (CH$_{PEG}$), 71.1–70.8 (CH$_{2\text{-PEG}}$), 70.2–69.9 (CH$_{PEG}$C$\underline{H}_2$O), 69.7 (OC$\underline{H}_2$CH$_2$), 64.4 (ArC$\underline{H}_2$O), 55.8 (O=SC$\underline{H}_2$C=O), 49.9 (O=SC$\underline{H}_2$CH$_2$), 23.2 (CH$_2$C$\underline{H}_2$CH$_2$) ppm; FTIR (spin coated film on a gold wafer): υ = 3243, 2924, 2870, 2375, 1742, 1653, 1613, 1578, 1530, 1447, 1374, 1346, 1276, 1121, 1057, 860 cm^{-1}.

6.3.4 Copolymers

Methoxy poly(ethylene glycol)-*block*-poly(allyl glycidyl ether) (mPEG-*b*-PAGE, 17–18)

The synthesis was performed according to procedure B (Chapter 6.3.1.2).

Polymer **17** was purified chromatographically to yield (60%) a white amorphous substance. R_f = 0.56 (chloroform/methanol, 10:1.5 v/v).

^{1}H NMR (500 MHz, CDCl$_3$): δ = 5.93–5.82 (m, 1H, C$\underline{H}$=CH$_2$), 5.25 (d, 1H, CH=C$\underline{H}_2$), 5.14 (d, 1H, CH=C$\underline{H}_2$), 3.98 (d, 2H, OC$\underline{H}_2$CH=CH$_2$), 3.80–3.42 (m, CH$_{2\text{-PEG}}$ + CH$_{\text{PEG}}$ + CH$_{\text{PEG}}$C$\underline{H}_2$O), 3.37 (s, 3H, CH$_3$O) ppm.

Poly(allyl glycidyl ether)-*block*-poly(ethylene glycol)-*block*-poly(allyl glycidyl ether) (PAGE-*b*-PEG-*b*-PAGE, 19–20)

The synthesis was performed according to procedure B (Chapter 6.3.1.2).

Polymer **19** was purified chromatographically to yield (70%) a colorless amorphous substance. R_f = 0.53 (chloroform/methanol, 10:1.5 v/v).

Polymer **20** was purified chromatographically to yield (76%) a white solid. R_f = 0.50 (chloroform/methanol, 10:1.5 v/v).

^{1}H NMR (500 MHz, CDCl$_3$): δ = 5.93–5.82 (m, 1H, C$\underline{H}$=CH$_2$), 5.25 (d, 1H, CH=C$\underline{H}_2$), 5.14 (d, 1H, CH=C$\underline{H}_2$), 3.98 (d, 2H, OC$\underline{H}_2$CH=CH$_2$), 3.80–3.42 (m, CH$_{2\text{-PEG}}$ + CH$_{\text{PEG}}$ + CH$_{\text{PEG}}$C$\underline{H}_2$O) ppm.

Methoxy poly(ethylene glycol)-*block*-poly(ethoxyethyl glycidyl ether) (mPEG-*b*-PEEGE, 21–22)

The synthesis was performed according to procedure B (Chapter 6.3.1.2).

Polymer **22** was purified chromatographically to yield (59%) a white solid. $R_f = 0.50$ (chloroform/methanol, 10:1.5 v/v).

^{1}H NMR (500 MHz, CDCl$_3$): δ = 4.73–4.67 (m, 1H, C$\underline{H}$CH$_3$), 3.80–3.40 (m, CH$_2$-PEG + CH$_{PEG}$ + CH$_{PEG}$C$\underline{H}_2$O + C$\underline{H}_2$CH$_3$), 3.38 (s, 3H, CH$_3$O), 1.31–1.27 (m, 3H, CHC$\underline{H}_3$), 1.19 (t, 3H, CH$_2$C$\underline{H}_3$) ppm.

Poly(ethoxyethyl glycidyl ether)-*block*-poly(ethylene glycol)-*block*-poly(ethoxyethyl glycidyl ether) (PEEGE-*b*-PEG-*b*-PEEGE, 23)

The synthesis was performed according to procedure B (Chapter 6.3.1.2).

^{1}H NMR (500 MHz, CDCl$_3$): δ = 4.73–4.67 (m, 1H, C$\underline{H}$CH$_3$), 3.80–3.40 (m, CH$_2$-PEG + CH$_{PEG}$ + CH$_{PEG}$C$\underline{H}_2$O + C$\underline{H}_2$CH$_3$), 1.31–1.27 (m, 3H, CHC$\underline{H}_3$), 1.19 (t, 3H, CH$_2$C$\underline{H}_3$) ppm.

4-Arms Star Poly(ethylene glycol)-*block*-poly(ethoxyethyl glycidyl ether) (C(PEG-*b*-PEEGE)₄, 24)

The synthesis was performed according to procedure B (Chapter 6.3.1.2).

^{1}H NMR (500 MHz, CDCl₃): δ = 4.73–4.67 (m, 1H, C$\underline{H}$CH₃), 3.80–3.40 (m, CH₂-PEG + CH$_{PEG}$ + CH$_{PEG}$C$\underline{H_2}$O + C$\underline{H_2}$CH₃), 1.31–1.27 (m, 3H, CHC$\underline{H_3}$), 1.19 (t, 3H, CH₂C$\underline{H_3}$) ppm.

Methoxy poly(ethylene glycol)-*block*-poly(glycidaldehyde diethyl acetal) (mPEG-*b*-PAGE, 25–26)

Block copolymer was synthesized by stepwise polymerization of EO (according to procedure A, Chapter 6.3.1.1) followed by polymerization of GADEA initiated by living chains of *in situ* prepared mPEG.

^{1}H NMR (500 MHz, CDCl₃): δ = 4.55–4.40 (m, 1H, C$\underline{H}$(OEt)₂), 3.95–3.40 (m, CH₂-PEG + CH$_{PEG}$ + 2 × C$\underline{H_2}$CH₃), 3.38 (s, 3H, CH₃O), 1.28–1.16 (CH₂C$\underline{H_3}$) ppm.

Triblock copolymer of poly(ethylene glycol) and photosensitive protected aldehyde poly(ethylene glycol) (PPAPEG-*b*-PEG-*b*-PPAPEG, 37)

The synthesis was performed according to procedure B (Chapter 6.3.1.2).

Polymer **37d** was purified chromatographically to yield (86%) white solid. $R_f = 0.53$ (chloroform/methanol, 10:1.5 v/v).

^{1}H NMR (500 MHz, CDCl$_3$): $\delta = 7.37$–7.20 (m, 10H, H$_{Ar}$), 6.92 and 6.88 (d, 1H, H$_{Ar}$), 6.78–6.74 (m, 1H, H$_{Ar}$), 6.35–6.33 (m, 1H, H$_{Ar}$), 5.01 and 4.96 (d, 1H, CH(O)$_2$), 4.08–4.02 (m, 1H, CH$_{PEG}$), 3.80–3.46 (m, CH$_{2\text{-PEG}}$ + CH$_3$OAr) ppm.

Copolymer of benzyl poly(ethylene glycol) and ester poly(ethylene glycol) (43)

Copolymer was synthesized via sequential reactions (according to procedure D, Chapter 6.3.1.4) of polymer **12b** (M$_n$ 6800 g/mol, 0.25 g, 2.2 mmol of allyl groups, 1 eq) with methyl mercaptoacetate (0.37 eq), followed by *in situ* reaction of the obtain polymer **42** with benzyl mercaptane (3.1 eq). The crude polymer was purified chromatographically on silica gel to yield 0.35 g (72%) of the title polymer (ester to benzyl groups ratio 1:1.7) as a colorless oil. $R_f = 0.80$ (chloroform/methanol, 10:0.7 v/v). GPC (THF): M$_n$ 9700 g/mol, M$_w$ 12200 g/mol, PDI 1.26. ^{1}H NMR (500 MHz, CDCl$_3$): $\delta = 7.34$–7.17 (5H, H$_{Ar}$), 3.70 (3H, CH$_3$OC=O), 3.67 (2H,

PhC<u>H</u>$_2$S), 3.64–3.32 (7H and 7H, CH$_2$-PEG + CH$_{PEG}$ + CH$_{PEG}$C<u>H</u>$_2$O + OC<u>H</u>$_2$CH$_2$), 3.35 (3H, CH$_3$O), 3.19 (2H, O=CCH$_2$S), 2.67 (2H, O=CCH$_2$SC<u>H</u>$_2$), 2.45 (2H, PhCH$_2$SC<u>H</u>$_2$), 1.82 (2H, O=CCH$_2$SCH$_2$C<u>H</u>$_2$), 1.78 (2H, PhCH$_2$SCH$_2$C<u>H</u>$_2$) ppm; ^{13}C NMR (125 MHz, CDCl$_3$): δ = 171.0 (C=O), 138.6 (C$_{Ar}$), 129.0 (2C, CH$_{Ar}$), 128.6 (2C, CH$_{Ar}$), 127.1 (CH$_{Ar}$), 79.1–78.6 (CH$_{PEG}$), 71.3–71.0 (CH$_2$-PEG), 70.3–70.1 (CH$_{PEG}$C<u>H</u>$_2$O), 70.0 and 69.7 (OC<u>H</u>$_2$CH$_2$), 52.5 (C<u>H</u>$_3$OC=O), 36.4 (SC<u>H</u>$_2$Ph), 33.5 (SC<u>H</u>$_2$C=O), 29.6 and 29.5 (SC<u>H</u>$_2$CH$_2$), 29.2 and 28.2 (CH$_2$C<u>H</u>$_2$CH$_2$) ppm.

Copolymer of benzyl poly(ethylene glycol) and hydrazide poly(ethylene glycol) (44)

The synthesis was performed according to procedure E (Chapter 6.3.1.5) from polymer **43** (M$_n$ 9700 g/mol, 0.3 g, 0.48 mmol of ester groups) resulting in 99.9% conversion of ester groups into hydrazide groups. The product was extracted with chloroform, washed with water, brine, dried over anhydrous sodium sulfate, filtered and concentrated *in vacuo* to yield 0.23 g (78%) of the title polymer (ester to benzyl groups ratio 1:1.7) as a yellow viscous oil. ^{1}H NMR (500 MHz, CDCl$_3$): δ = 8.2 (1H, NH$_2$N<u>H</u>), 7.34–7.17 (5H, H$_{Ar}$), 3.92 (2H, N<u>H</u>$_2$NH), 3.67 (2H, PhC<u>H</u>$_2$S), 3.64–3.32 (7H and 7H, CH$_2$-PEG + CH$_{PEG}$ + CH$_{PEG}$C<u>H</u>$_2$O + OC<u>H</u>$_2$CH$_2$), 3.35 (3H, CH$_3$O), 3.19 (2H, O=CCH$_2$S), 2.63 (2H, O=CCH$_2$SC<u>H</u>$_2$), 2.45 (2H, PhCH$_2$SC<u>H</u>$_2$), 1.82 (2H, O=CCH$_2$SCH$_2$C<u>H</u>$_2$), 1.78 (2H, PhCH$_2$SCH$_2$C<u>H</u>$_2$) ppm; ^{13}C NMR (125 MHz, CDCl$_3$): δ = 169.6 (C=O), 138.6 (C$_{Ar}$), 129.0 (2C, CH$_{Ar}$), 128.6 (2C, CH$_{Ar}$), 127.1 (CH$_{Ar}$), 79.1–78.6 (CH$_{PEG}$), 71.3–71.0 (CH$_2$-PEG), 70.3–70.1 (CH$_{PEG}$C<u>H</u>$_2$O), 70.0 and 69.7 (OC<u>H</u>$_2$CH$_2$), 36.4 (SC<u>H</u>$_2$Ph), 33.5 (SC<u>H</u>$_2$C=O), 29.6 and 29.5 (SC<u>H</u>$_2$CH$_2$), 29.2 and 28.2 (CH$_2$C<u>H</u>$_2$CH$_2$) ppm.

2-Methoxyethoxy poly(allyl glycidyl ether-*co*-benzyl glycidyl ether) (P(AGE-*co*-BnGE, 45)

The synthesis was performed according to procedure B (Chapter 6.3.1.2) with 2-methoxyethanol:AGE:BnGE ratio 1:20:40 resulting in 99% AGE conversion and 97% BnGE conversion. The crude polymer was purified chromatographically on silica gel to yield (72%) the title polymer (AGE to BnGE ratio 1:2) as a slightly yellow oil. R_f = 0.68 (chloroform/methanol, 10:0.7 v/v). GPC (THF): M_n 7600 g/mol, M_w 8000 g/mol, PDI 1.06. ^{1}H NMR (500 MHz, CDCl$_3$): δ = 7.34–7.18 (5H, H$_{Ar}$), 5.83 (1H, C$\underline{H}$=CH$_2$), 5.16 (2H, CH=C$\underline{H}_2$), 4.46 (2H, PhCH$_2$O), 3.92 (2H, OC$\underline{H}_2$CH=CH$_2$), 3.75–3.37 (5H and 5H, CH$_{2\text{-PEG}}$ + CH$_{PEG}$ + CH$_{PEG}$C$\underline{H}_2$O), 3.33 (3H, CH$_3$O) ppm; ^{13}C NMR (125 MHz, CDCl$_3$): δ = 138.6 (C$_{Ar}$), 135.1 (CH$_2$=$\underline{C}$H), 128.4 (2C, CH$_{Ar}$), 127.7 (2C, CH$_{Ar}$), 127.6 (CH$_{Ar}$), 116.8 ($\underline{C}$H$_2$=CH), 79.1–78.8(CH$_{PEG}$), 73.4 (O$\underline{C}$H$_2$Ph), 72.3 (O$\underline{C}$H$_2$CH=CH$_2$), 70.5–69.8 (2C, CH$_{2\text{-PEG}}$ + CH$_{PEG}$C$\underline{H}_2$O), 59.1(CH$_3$O) ppm.

Copolymer of poly(benzyl glycidyl ether) and ester poly(ethylene glycol) (46)

The synthesis was performed according to procedure D (Chapter 6.3.1.4) from polymer **45** (M$_n$ 7600 g/mol, 0.4 g, 0.9 mmol of allyl groups) resulting in 99.9% conversion of allyl groups into ester groups. The crude polymer was purified chromatographically on silica gel to yield 0.4 g (81%) of the title polymer as a slightly yellow oil. R_f = 0.70 (chloroform/methanol, 10:0.7 v/v). GPC (THF): M_n 8300 g/mol, M_w 9100 g/mol, PDI 1.10. ^{1}H NMR (500 MHz, CDCl$_3$): δ = 7.35–7.18 (5H, H$_{Ar}$), 4.46 (2H, PhCH$_2$O), 3.69 (3H, CH$_3$), 3.66–3.36 (7H and 5H, CH$_{2\text{-PEG}}$ + CH$_{PEG}$ + CH$_{PEG}$C$\underline{H}_2$O + OC$\underline{H}_2$CH$_2$), 3.37 (3H, CH$_3$O), 3.17 (2H, SCH$_2$C=O), 2.64 (2H, SC$\underline{H}_2$CH$_2$), 1.80 (2H, CH$_2$C$\underline{H}_2$CH$_2$) ppm; ^{13}C NMR (125 MHz, CDCl$_3$): δ = 171.0

(C=O), 138.6 (C$_{Ar}$), 128.4 (2C, CH$_{Ar}$), 127.6 (3C, CH$_{Ar}$), 79.1–78.7 (CH$_{PEG}$), 73.4 (O$\underline{C}$H$_2$Ph), 71.1–68.8 (2C, CH$_{2\text{-PEG}}$ + CH$_{PEG}$$\underline{C}H_2$O), 69.7 (O$\underline{C}H_2CH_2$), 59.1 (CH$_3$O), 52.5 ($\underline{C}H_3$OC=O), 33.5 (S$\underline{C}H_2$C=O), 29.5 (S$\underline{C}H_2CH_2$), 29.2 (CH$_2$$\underline{C}H_2CH_2$) ppm.

Copolymer of poly(benzyl glycidyl ether) and hydrazide poly(ethylene glycol) (P(HZ-*co*-BnGE), 47)

The synthesis was performed according to procedure E (Chapter 6.3.1.5) from polymer **46** (M$_n$ 8300 g/mol g/mol, 0.29 g, 0.5 mmol of ester groups) resulting in 99.9% conversion of ester groups into hydrazide groups. The product was extracted with chloroform, washed with water, brine, dried over anhydrous sodium sulfate, filtered and concentrated *in vacuo* to yield 0.20 g (71%) of the title polymer (hydrazide to benzyl groups ratio 1:2) as a slightly yellow viscous oil. ^{1}H NMR (500 MHz, CDCl$_3$): δ = 7.35–7.18 (5H, H$_{Ar}$), 4.47 (2H, PhCH$_2$O), 3.80–3.30 (7H and 5H, CH$_{2\text{-PEG}}$ + CH$_{PEG}$ + CH$_{PEG}$$\underline{C}H_2$O + O$\underline{C}H_2CH_2$), 3.34 (3H, CH$_3$O), 3.16 (2H, SCH$_2$C=O), 2.60 (2H, S$\underline{C}H_2CH_2$), 1.78 (2H, CH$_2$$\underline{C}H_2CH_2$) ppm; ^{13}C NMR (125 MHz, CDCl$_3$): δ = 169.9 (C=O), 138.5 (C$_{Ar}$), 128.5 (2C, CH$_{Ar}$), 127.7 (3C, CH$_{Ar}$), 79.1–78.7 (CH$_{PEG}$), 73.4 (O$\underline{C}$H$_2$Ph), 70.5–69.5 (2C, CH$_{2\text{-PEG}}$ + CH$_{PEG}$$\underline{C}H_2$O), 62.6 (O$\underline{C}H_2CH_2$), 34.4 (S$\underline{C}H_2$C=O), 29.8 (S$\underline{C}H_2CH_2$), 29.2 (CH$_2$$\underline{C}H_2CH_2$) ppm.

6.4   Surface Modification

6.4.1   Immobilization of Epoxide

Microcontact Printing (General Procedure F)

Polydimethylsiloxane (PDMS) stamp (featured or flat) was covered with the solution of epoxide **27** (65 mg/mL in absolute ethanol) and left to dry in the air. After complete evaporation of solvent, the stamp was applied to the amino functionalized substrate for 1 h. Modified surfaces were thoroughly washed with ethanol to remove unreacted epoxide and dried under nitrogen flow.

Incubation in Solution

Amino functionalized surfaces were incubated in solution of epoxide **27** (65 mg/mL in absolute ethanol) overnight. Modified surfaces were thoroughly washed with ethanol to remove unreacted epoxide and dried under nitrogen flow.

6.4.2   Immobilization of Alexa Fluor 555

Azide conjugated red fluorescent dye Alexa Fluor 555 (1 mg/mL in absolute ethanol) was microcontact printed on the alkyne modified surfaces according to procedure F. After washing and drying the samples were imaged by fluorescent microscopy.

6.4.3   Immobilization of Biotine-Azide

Biotin-azide (500 µg/mL in absolute ethanol/water, 1:1 v/v) was microcontact printed on the alkyne modified surfaces according to procedure F. Modified surfaces were additionally washed with water and dried.

6.4.4   Conjugation of RB-Streptavidin

Biotin modified surfaces were incubated in solution of RB-Streptavidin (50 µg/mL in wash buffer solution: phosphate buffered saline (PBS) containing 0.02 v/v% Tween 20 and 0.1 w/v% bovine serum albumin (BSA)) for 2 h. Modified surfaces were thoroughly washed with wash buffer solution and water, dried under nitrogen flow and imaged by fluorescence microscopy.

6.5   Photo Induced Reactions on a Surface

6.5.1   Cleavage of Acetal Protecting Groups in PPAPEG

The samples were prepared by spin coating of PPAPEG (**33**) solution (20 mg/mL in dry THF) on gold wafers at approximately 100 rps for 30 sec using *Lot-Oriel* SCV-20 spin coater. The prepared samples were exposed to UV light (*Bio-Link* BLX 254 irradiation system with 5 × 8 W lamps, 312 nm) for different periods of time and analyzed by IR spectroscopy using *Bruker* Vertex 80 spectrometer.

6.5.2   Cleavage of NB Protecting Groups in ONBPEG

The samples were prepared by spin coating of ONBPEG (**52**) solution (50 mg/mL in chloroform) on gold wafers at approximately 83 rps for 30 sec using *Laurel* WS-400-6NPP spin coater. The prepared samples were exposed to UV light (*Maxima* ML-3500C ultra-high intensity UV-A Lamp, 120 W lamp, 365 nm) for different periods of time and analyzed by IR spectroscopy using *Thermo* Nicolet 6700 spectrometer.

6.6   Photo Induced Reactions in Solution

6.6.1   Cleavage of Photosensitive Acetal Protection Combined with *in situ* Reaction with Hydrazide

6.6.1.1   General Procedure G

A compound bearing PPA group(s) and a compound bearing hydrazide group(s) were dissolved in a corresponding solvent chosen according to the reagents solubility. The prepared solution was placed in a quartz cuvette (0.5 or 1 cm) and exposed to UV light (*Bio-Link* BLX 254 irradiation system with 5 × 8 W lamps, 312 nm) for a defined period of time while stirring.

Hydrazone 36

Hydrazone **36** (as a mixture of isomers 1.7:1) was prepared according to procedure G from diol **34-D2** (10 mg, 0.023 mmol, 1 eq) and acetohydrazide (10 mg, 0.135 mmol, 5.9 eq) in 1 mL of acetonitrile-d_3/deuterated water (9:1 v/v) solution by irradiation for 2 h and analyzed by ^{1}H NMR spectroscopy.

Hydrazone 39

Hydrazone **39** was prepared according to procedure G from PPAPEG-*b*-PEG-*b*-PPAPEG (**37d**, M_n 4700 g/mol, 20 mg, ~ 0.011 mmol of PPA groups, 1 eq) and acetohydrazide (10 mg, 0.135 mmol, 13.5 eq) in 1 mL of acetonitrile-d_3/deuterated water (9:1 v/v) solution by irradiation for 30 min (not complete conversion) or 2 h (complete conversion) and analyzed by ^{1}H NMR spectroscopy.

6.6.1.2   Hydrogel Fabrication

The hydrogel was prepared according to procedure G from PPAPEG-*b*-PEG-*b*-PPAPEG (**37d**, M_n 4700 g/mol, 45 mg, ~ 0.024 mmol of PPA groups, 1 eq) and PHZ (**41**, M_n ~ 18900 g/mol, 6.75 mg, ~ 0.03 mmol of hydrazide groups, 1 eq) in 0.52 mL of distilled water by irradiation for 2 h (stirring was performed for the first 30 min) and washed with distilled water.

6.6.2   Cleavage of NB Protecting Groups in ONBPEG

5 mg/mL solution of ONBPEG (**52**) in DMSO-d_6 was placed in a quartz cuvette (1 cm) and exposed to UV light (*Bio-Link* BLX 254 irradiation system with 5 × 8 W lamps, 312 nm) for different periods of time while stirring. The degree of the deprotection of NB groups was estimated by ^{1}H NMR spectroscopy.

.

6.7   Fabrication and Modification of Microparticles

6.7.1   General Procedures

6.7.1.1   General Procedure H for Fabrication of Bicompartmental Microspheres

Particles were fabricated by the EHD co-jetting process. Two different polymer solutions were prepared in separate vials. One solution was composed entirely of PLGA (40,000 g/mol), another – of a mixture of PLGA and a functional PEG. Mixture of chloroform and dimethylformamide (DMF) (95:5 v/v) was used as a solvent. The experimental setup of EHD co-jetting included a syringe pump (*Fisher Scientific*), a power supply (DC voltage source, *Gamma High Voltage Research*), and a flat grounded collector. Each one of the two polymer solutions was delivered at a constant flow rate of 0.4 mL/h via vertically positioned side-by-side syringes equipped with 26 G needles (*Hamilton*). When a driving voltage of 6–7 kV was applied to the polymer solutions, a stable Taylor Cone was formed and microspheres were collected at a distance of 30–40 cm. The jetted particles were allowed to dry *in vacuo* for one week in order to evaporate any remaining solvents. The particles were collected from the metal substrate using a buffer composed of PBS, 0.01% Tween 20, and 10% BSA. Once collected, the particles were filtered through a 40 µm filter to remove large aggregates and then were counted using a cell counter to determine the final concentration in the solution.

6.7.1.2   General Procedure I for Fabrication of Bicompartmental Microfibers

In the case of microfibers, the experimental setup was similar to that of the microspheres except the concentration of polymers in both solutions was increased and PLGA (55,000–75,000 g/mol) was used. Each one of the two polymer solutions was delivered at a constant flow rate of 0.05 mL/h with a driving voltage of 12 kV. The microfibers were collected at a distance of 7 cm using a rotary collector.

6.7.1.3   General Procedure J for Fabrication of Bicompartmental Microcylinders

Once the bicompartmental microfibers were fabricated, the cylinders were produced by cryosectioning as previously reported.[59] Briefly, the microfibers were embedded in a freezing medium (OCT) and sectioned at –20°C by a cryostat microtome (HM550 OMC, Microme) with a desired length of 70 um. After the samples were collected, remaining OCT was removed by extensive washing with 0.01 v/v% Tween 20/DI-water at room temperature.

6.7.2   Fabrication and Modification of PLGA/P(HZ-*co*-BnGE) Microparticles

6.7.2.1   Fabrication of PLGA/P(HZ-*co*-BnGE) Microspheres

The microspheres were generated according to procedure H (Chapter 6.7.1.1) with the following jetting solutions: one solution contained 9 w/v% PLGA, another solution was prepared by mixing 9 w/v% of PLGA with 0.9 w/v% of P(HZ-*co*-BnGE) (10% by weight of PLGA). Microspheres were collected at a distance of 40 cm.

6.7.2.2   Fabrication of PLGA/P(HZ-*co*-BnGE) Microfibers

The microfibers were generated according to procedure I (Chapter 6.7.1.2) with the following jetting solutions: one solution contained 30 w/v% PLGA, another solution was prepared by mixing 30 w/v% of PLGA with 15 w/v% of P(HZ-*co*-BnGE) (50% by weight of PLGA).

6.7.2.3   Fabrication of PLGA/P(HZ-*co*-BnGE) Microcylinder

Bicompartmental PLGA/P(HZ-*co*-BnGE) microcylinders were fabricated according to procedure J (Chapter 6.7.1.3).

6.7.2.4   Selective Fluorescence-Labeling of PLGA/P(HZ-*co*-BnGE) Microspheres

PLGA/P(HZ-*co*-BnGE) microspheres were first collected and suspended in an aqueous solution containing 0.01% Tween 20. To visually confirm the presence of hydrazide groups on the particle surface, rhodamine-labeled carboxyl groups (RB-PEG-COOH) were conjugated using EDC/NHS chemistry. Briefly, 10 mM RB-PEG-COOH/PBS containing 0.01% Tween 20 in a total volume of 1 mL was activated with 40 mM EDC for 10 min, and then with 10 mM sulfo-NHS for another 10 min. Next, approximately 300 µg of the particles were reacted with the solution for 2 hours. Finally, the particles were washed 20 times by centrifugation to remove any unreacted chemicals, and the selective binding of rhodamine on the particles was confirmed through CLSM.

6.7.2.5   Selective Sugar-Lectin Binding of PLGA/P(HZ-*co*-BnGE) Microcylinders

Prior to the sugar-lectin reaction, 2α-mannobiose was first oxidized with sodium periodate to obtain free aldehyde groups. The aldehyde groups were then selectively conjugated with hydrazide groups on the cylinder surface. The reaction condition was as follows. Approximately 50,000 bicompartmental PLGA/P(HZ-*co*-BnGE) microcylinders were kept rotating with 0.85 mg of 2α-mannobiose and 2.14 mg of sodium periodate in Eppendorf tube containing 1 mL of 0.01 v/v% Tween 20/DI-water for 5 h at room temperature. The unreacted molecules were removed by repeated washing with 0.01 v/v% Tween 20/DI-water. In order to confirm selective immobilization of 2α-mannobiose on the microcylinders, a sugar-lectin reaction was performed by incubating the mannose-binding lectin Con A. The 2α-mannobiose-bound microcylinders were reacted with 100 µg of TRITC-Con A and gently rotated in the buffer solution (10 mM HEPES, 1 mM CaCl$_2$, 1 mM MnCl$_2$, 0.01% Tween 20) for 3 h at room temperature. The unreacted molecules were removed by washing with the same buffer solution.

6.7.3 Fabrication and Modification of PLGA/NBPEG and PLGA/ONBPEG Microparticles

6.7.3.1 Fabrication of PLGA/NBPEG and PLGA/ONBPEG Microspheres

PLGA/NBPEG and PLGA/ONBPEG microspheres were generated according to procedure H (Chapter 6.7.1.1) with the following jetting solutions: one solution contained 15 w/v% PLGA, another solution contained 15 w/v% of 1:1 w/w mixture of PLGA and NBPEG(ONBPEG). Microspheres were collected at a distance of 30 cm.

For confocal Raman microscopy analysis of PLGA/ONBPEG microspheres the ratio of PLGA and ONBPEG in the second solution was changed to 3:1 w/w correspondingly.

6.7.3.2 Fabrication of PLGA/NBPEG and PLGA/ONBPEG Microfibers

PLGA/NBPEG and PLGA/ONBPEG microfibers were generated according to procedure I (Chapter 6.7.1.2) with the following jetting solutions: one solution contained 30 w/v% PLGA, another solution was prepared by mixing 30 w/v% of PLGA with 22.5 w/v% of NBPEG(ONBPEG) (75% by weight of PLGA).

For confocal Raman microscopy analysis of PLGA/ONBPEG microfibers the concentration of ONBPEG in the second solution was decreased to 15 w/v% (50% by weight of PLGA).

6.7.3.3 Photodegradation of PLGA/ONBPEG Microparticles

PLGA/NBPEG and PLGA/ONBPEG microspheres or microfibers were exposed to UV light (*Maxima* ML-3500C ultra-high intensity UV-A Lamp, 120 W lamps, 365 nm) on the collecting substrate for 30 min, followed by incubation in a buffer composed of tris-buffered saline (TBS) and 0.01% Tween 20 for 1 h while rotating. Next, the particles were washed 5 times with DI-water, dried *in vacuo* overnight and analyzed by SEM.

As a control experiment, incubation of non-irradiated particles in the same conditions was performed.

6.7.3.4 *In Vitro* Incubation and Assessment

Raw264.7 cells were seeded in a 12-well plate with cover glass slides at a concentration of ~80000 cells/well. The following day, the cells were incubated with particles at different concentrations (in fresh media) for four hours. After the incubation, the cells were washed with PBS, and fixed in 4% paraformaldehyde. Following fixation, the cells were washed with PBS once more and stained with far-red phalloidin, as per the protocol provided by *Invitrogen*. The coverslip samples were then mounted on glass slides using *ProLong* Gold Antifade Reagent with DAPI and imaged through CLSM. The scanned images were then analyzed to determine particle uptake per cell at the different particle concentrations.

7   List of Abbreviations

Ac	acetyl
AG	aminoglycidol
AGE	allyl glycidyl ether
AIBN	azobisisobutyronitrile
ATR	Attenuated Total Reflection
ATRP	atom transfer radical polymerization
Bn	benzyl
BnGE	benzyl glycidyl ether
Bp	boiling point
BSA	bovine serum albumin
BtOH	1-hydroxybenzotriazole
Bu	butyl
t-Bu	$tert$-butyl
t-BuGE	$tert$-butyl glycidyl ether
Con A	concanavalin A
CLSM	Confocal Laser Scanning Microscopy
cm	centimeter
CMC	critical micelle concentration
mCPBA	$meta$-chloroperoxybenzoic acid
CVD	chemical vapor deposition
conc.	concentrated
δ	chemical shift
DAAG	N,N-diallylaminoglycidol
DBAG	N,N-dibenzyl amino glycidol
DEPT	Distortionless Enhancement by Polarization Transfer
DI-water	deionized water
DIC	N,N'-diisopropylcarbodiimide
DMAc	N,N-dimethylacetamide
DMAP	4-dimethylaminopyridine
DMF	dimethylformamide
DMSO	dimethyl sulfoxide
DOPA	3,4-dihydroxyphenylalanine

DP_n	degree of polymerization
DPMK	diphenylmethylpotassium
EA	Elemental Analysis
ECH	epichlorohydrine
ECM	extracellular matrix
EDC	N-(3-dimethylaminopropyl)-N'-ethylcarbodiimide
EEGE	ethoxyethyl glycidyl ether
EHD	electrohydrodynamic
EI	Electron Ionization
EO	ethylene oxide
eq	equivalent
Et	ethyl
eV	electron volt
EVGE	ethoxy vinyl glycidyl ether
FDA	Food and Drug Administration
FEG-SEM	Field Emission Scanning Electron Microscope
FGE	furfuryl glycidyl ether
FO	2-furyloxyrane
FT	Fourier Transformation
g	gram
GADEA	glycidaldehyde diethyl acetal
GPC	Gel Permeation Chromatography
H	hour
HEPES	4-(2-hydroxyethyl)-1-piperazineethanesulfonic acid
Hz	hertz
HZ	hydrazide
IGG	1,2-isopropylidene glyceryl glycidyl ether
IR	infrared
J	coupling constants
KHMDS	potassium bis(trimethylsilyl)amide
kV	kilovolt
kW	kilowatt
M	molar

MALDI-TOF-MS	Matrix-Assisted Laser Desorption/Ionization – Time Of Flight Mass Spectrometry
mbar	millibar
Me	methyl
Mes	mesyl (methanesulfonyl)
μg	microgram
mg	milligram
MHz	megahertz
min	minute
mL	milliliter
μm	micrometer
mM	mollimolar
mmol	millimol
M_n	number average molecular weight
Mp	melting point
mPEG	methoxy poly(ethylene glycol)
MS	Mass Spectrometry
M_w	weight average molecular weight
NB	*o*-nitrobenzyl
NBPEG	*o*-nitrobenzyl poly(ethylene glycol)
NHS	N-hydroxysuccinimide
nm	nanometer
NMR	Nuclear Magnetic Resonance
OEGMA	oligo(ethylene glycol) methyl methacrylate
ONBPEG	oxidized *o*-nitrobenzyl poly(ethylene glycol)
PAA	poly(acrylic acid)
PDAAG	poly(N,N-diallylaminoglycidol)
PAG	poly(aminoglycidol)
PAGE	poly(allyl glycidyl ether)
PBnGE	poly(benzyl glycidyl ether)
PBO	poly(butylene oxide)
PBS	phosphate buffered saline
PDBAG	poly(N,N-dibenzyl amino glycidol)
PDI	polydispersity index

PDMS	polydimethylsiloxane
PEG	poly(ethylene glycol)
PECH	poly(epichlorohydrine)
PEEGE	poly(ethoxyethyl glycidyl ether)
PEI	polyethylene imine
PEVGE	poly(ethoxy vinyl glycidyl ether)
PEO	poly(ethylene oxide)
PFO	poly(2-furyloxyrane)
PG	polyglycidol
PGA	poly(glycolic acid)
PGADEA	poly(glycidaldehyde diethyl acetal)
Ph	phenyl
PhGE	phenyl glycidyl ether
PHO	poly(hexylene oxide)
PHZ	hydrazide poly(ethylene glycol)
PIGG	1,2-isopropylidene glyceryl glycidyl ether
PLA	poly(lactic acid)
PLGA	poly(lactic-*co*-glycolic acid)
PN	potassium naphthalenide
PO	propylene oxide
POO	poly(octylene oxide)
PPA	photolabile protected aldehyde
PPAEG	photosensitive protected aldehyde ethylene glycol
PPAPEG	photosensitive protected aldehyde poly(ethylene glycol)
ppm	part per million
PPO	poly(propylene oxide)
PPS	poly(propylene sulfide)
i-Pr	isopropyl
PS	polystyrene
RB-steptavidin	rhodamine B streptavidin
RGD	tripeptide L-arginyl – glycyl – L-aspartic acid
R_f	retention factor
rps	rounds per second
rt	room temperature

SAM	self-assembled monolayer
sec	second
SEM	Scanning Electron Microscopy
Sulfo-NHS	N-hydroxysulfosuccinimide
TBS	Tris-buffered saline
THF	tetrahydrofuran
THP	tetrahydropyranyl
TLC	thin layer chromatography
TMS	trimethylsilyl
Ts	tosyl (p-toluenesulfonyl)
UV	ultraviolet
UV-Vis	Ultraviolet-Visible Spectroscopy
υ	wavenumber
V	volt
v/v	volume/volume ratio
v/v%	volume/volume percent
W	watt
w/w	weight/weight ratio
w/v%	weight/volume percent
XPS	X-ray photoelectron spectroscopy

8   References

[1]   R. Ravichandran, S. Sundarrajan, J. R. Venugopal, Sh. Mukherjee, S. Ramakrishna, *Macromol. Biosci.* **2012**, *12*, 286–311. *Advances in polymeric systems for tissue engineering and biomedical applications.*

[2]   Y. Wang, J. D. Byrne, M. E. Napier, J. M. De Simone, *Adv. Drug Deliv. Rev.* **2012**, *64*, 1021–1030. *Engineering nanomedicines using stimuli-responsive biomaterials.*

[3]   M. M. Stevens, G. Mecklenburg, *Polym. Int.* **2012**, *61*, 680–685. *Bio-inspired materials for biosensing and tissue engineering.*

[4]   V. K. Vendra, L. Wu, S. Krishnan. In Nanomaterials for the Life Sciences. Vol.5: Nanostructured Thin Films and Surfaces; Kumar Ch., Ed.; WILEY-VCH Verlag GmbH & Co. KGaA: Weinheim, Germany, 2010; Chapter 1, pp. 1–54.

[5]   V. Hasirci, E. Vrana, P. Zorlutuna, A. Ndreu, P. Yilgor, F. B. Basmanav, E. Aydin, *J. Biomater. Sci. Polym. Ed.* **2006**, *17*, 1241–1268. *Nanobiomaterials: A review of the existing science and technology, and new approaches.*

[6]   M. Yoshida, J. Lahann, *ACS Nano* **2008**, *2*, 1101–1107. *Smart nanomaterials.*

[7]   I. Tomatsu, K. Peng, A. Kros, *Adv. Drug Deliv. Rev.* **2011**, *63*, 1257–1266. *Photoresponsive hydrogels for biomedical applications.*

[8]   Y. Y. Li, H. Q. Dong, K. Wang, D. L. Shi, X. Zh. Zhang, R. X. Zhuo, *Sci. China Chem.* **2010**, *53*, 447–457. *Stimulus-responsive polymeric nanoparticles for biomedical applications.*

[9]   M. A. Cohen Stuart, W. T. S. Huck, J. Genzer, M. Müller, Ch. Ober, M. Stamm, G. B. Sukhorukov, I. Szleifer, V. V. Tsukruk, M. Urban, F. Winnik, S. Zauscher, I. Luzinov, S. Minko, *Nature Mater.* **2010**, *9*, 101–113. *Emerging applications of stimuli-responsive polymer materials.*

[10]   H. Sashiwa, S. Aiba, *Prog. Polym. Sci.* **2004**, *29*, 887–908. *Chemically modified chitin and chitosan as biomaterials.*

[11] J. D. Kosmala, D. B. Henthorn, L. Brannon-Peppas, *Biomaterials* **2000**, *21*, 2019–2023. *Preparation of interpenetrating networks of gelatin and dextran as degradable biomaterials.*

[12] L. Pescosolido, W. Schuurman, J. Malda, P. Matricardi, F. Alhaique, T. Coviello, P. R. van Weeren, W. J. A. Dhert, W. E. Hennink, T. Vermonden. *Biomacromolecules* **2011**, *12*, 1831–1838. *Hyaluronic acid and dextran-based semi-IPN hydrogels as biomaterials for bioprinting.*

[13] B. D. Ulery, L. S. Nair, C. T. Laurencin, *J. Polym. Sci., Part B: Polym. Phys.* **2011**, *49*, 832–864. *Biomedical applications of biodegradable polymers.*

[14] E. Fournier, C. Passirani, C. N. Montero-Menei, J. P. Benoit, *Biomaterials* **2003**, *24*, 3311–3331. *Biocompatibility of implantable synthetic polymeric drug carriers: Focus on brain biocompatibility.*

[15] O. D. Krishna, K. L. Kiick, *Biopolymers* **2010**, *94*, 32–48. *Protein- and peptide-modified synthetic polymeric biomaterials.*

[16] G. Pasut, F.M. Veronese, *Adv. Drug Deliver. Rev.* **2009**, *61*, 1177–1188. *PEG conjugates in clinical development or use as anticancer agents: An overview.*

[17] U. Wattendorf, H. P. Merkle, *J. Pharm. Sci.* **2008**, 97, 4655–4669. *PEGylation as a tool for the biomedical engineering of surface modified microparticles.*

[18] F. Zhang, E. T. Kang, K. G. Neoh, P. Wang, K. L. Tan, *Biomaterials* **2001**, *22*, 1541–1548. *Surface modifcation of stainless steel by grafting of poly(ethylene glycol) for reduction in protein adsorption.*

[19] K. B. Bjugstad, K. Lampe, D. S. Kern, M. Mahoney, *J. Biomed. Mater. Res. Part A* **2010**, *95*, 79–91. *Biocompatibility of poly(ethylene glycol)-based hydrogels in the brain: An analysis of the glial response across space and time.*

[20] B. Obermeier, F. Wurm, Ch. Mangold, H. Frey, *Angew. Chem. Int. Ed.* **2011**, *50*, 7988–7997. *Multifunctional poly(ethylene glycol)s.*

[21]   M. E. Davis, Zh. Chen, Dong. M. Shin, *Nat. Rev. Drug Discov.* **2008**, *7*, 771–782. *Nanoparticle therapeutics: An emerging treatment modality for cancer.*

[22]   G. Manivasagam, D. Dhinasekaran, A. Rajamanickam, *Recent Patents Corros. Sci.,* **2010**, *2*, 40-54. *Biomedical implants: Corrosion and its prevention.*

[23]   P. K. Chu, *Thin Solid Films* **2013**, *528*, 93–105. *Surface engineering and modification of biomaterials.*

[24]   J. H. Lee, H. B. Lee, J. D. Andrades, *Prog. Polym. Sci.* **1995**, *20*, 1043–1079. *Blood compatibility of polyethylene oxide surfaces.*

[25]   R. G. Chapman, E. Ostuni, M. N. Liang, G. Meluleni, E. Kim, L. Yan, G. Pier, H. Sh. Warren, G. M. Whitesides, *Langmuir* **2001**, *17*, 1225–1233. *Polymeric thin films that resist the adsorption of proteins and the adhesion of bacteria.*

[26]   G. Lapienis, *Prog. Polym. Sci.* **2009**, *34*, 852–892. *Star-shaped polymers having PEO arms.*

[27]   A. M. Carmona-Ribeiro, L. Barbassa, L. Dias de Melo. In Biomimetic Based Applications; M. Cavrak, Ed.; InTech Europe: Rijeka, Croatia, 2011; Chapter 10, pp. 229–284.

[28]   J. Groll, T. Ameringer, J. P. Spatz, M. Möller, *Langmuir* **2005**, *21*, 1991–1999. *Ultrathin coatings from isocyanate-terminated star PEG prepolymers: Layer formation and characterization.*

[29]   P. Gasteier, A. Reska, P. Schulte, J. Salber, A. Offenhäusser, M. Möller, J. Groll, *Macromol. Biosci.* **2007**, *7*, 1010–1023. *Surface grafting of PEO-based star-shaped molecules for bioanalytical and biomedical applications.*

[30]   N. Herzer, S. Hoeppener, U. S. Schubert, *Chem. Commun.* **2010**, *46*, 5634–5652. *Fabrication of patterned silane based self-assembled monolayers by photolithography and surface reactions on silicon-oxide substrates.*

[31]   B. Brough, K. L. Christman, T. S. Wong, Ch. M. Kolodziej, J. G. Forbes, K. Wang, H.
       D. Maynard, Ch.-M. Ho, *Soft Matter.* **2007**, *3*, 541–546. *Surface initiated actin
       polymerization from top-down manufactured nanopatterns.*

[32]   J. Rundqvist, J. H. Hoh, D. B. Haviland, *Langmuir* **2005**, *21*, 2981–2987.
       Poly(ethylene glycol) self-assembled monolayer island growth.

[33]   F. Meiners, I. Plettenberg, J. Witt, B. Vaske, A. Lesch, I. Brand, G. Wittstock, *Anal.
       Bioanal. Chem.* **2013**, *405*, 3673–3691. *Local control of protein binding and cell
       adhesion by patterned organic thin films.*

[34]   H. Ma, J. Hyun, P. Stiller, A. Chilkoti, *Adv. Mater.* **2004**, *16*, 338–341. *"Non-fouling"
       oligo(ethylene glycol)-functionalized polymer brushes synthesized by surface-initiated
       atom transfer radical polymerization.*

[35]   J. L. Dalsin, L. Lin, S. Tosatti, J. Vörös, M. Textor, Ph. B. Messersmith, *Langmuir*
       2005, *21*, 640–646. *Protein resistance of titanium oxide surfaces modified by
       biologically inspired mPEG-DOPA.*

[36]   S. Zürcher, D. Wäckerlin, Y. Bethuel, B. Malisova, M. Textor, S. Tosatti, K.
       Gademann, *J. Am. Chem. Soc.* **2006**, *128*, 1064–1065. *Biomimetic surface
       modifications based on the cyanobacterial iron chelator anachelin.*

[37]   J. Groll, Th. Ameringer, J. P. Spatz, M. Möller, *Langmuir* **2005**, *21*, 1991–1999.
       *Ultrathin coatings from isocyanate-terminated star peg prepolymers: Layer formation
       and characterization.*

[38]   M.L. Graham, *Adv. Drug Deliver. Rev.* **2003**, *55*, 1293–1302. *Pegaspargase: A review
       of clinical studies.*

[39]   G. Molineux, *Curr. Pharm. Des.* **2004**, *10*, 1235–1244. *The design and development of
       Pegfilgrastim (PEG-rmetHuG-CSF, Neulasta).*

[40]   Y.-S. Wang, S. Youngster, M. Grace, J. Bausch, R. Bordens, D. F. Wyss, *Adv. Drug
       Deliver. Rev.* **2002**, *54*, 547–570. *Structural and biological characterization of
       pegylated recombinant interferon alpha-2b and its therapeutic implications.*

[41]   T. M. Herndon, S. G. Demko, X. Jiang, K. He, J. E. Gootenberg, M. H. Cohen, P. Keegan, R. Pazdur, *Oncologist* **2012**, *17*, 1323–1328. *U.S. Food and Drug Administration approval: Peginterferon-alfa-2b for the adjuvant treatment of patients with melanoma.*

[42]   Sh. S. Banerjee, N. Aher, R. Patil, J. Khandare, *J. Drug Deliv.* **2012**, *2012*, Article ID, 103973, 17 pp. *Poly(ethylene glycol)-prodrug conjugates: Concept, design, and applications.*

[43]   K. Mondon, R. Gurny, M. Möller, Chimia **2008**, *62*, 832–840. *Colloidal drug delivery systems – recent advances with polymeric micelles.*

[44]   D. A. Chiappetta, A. Sosnik, *Eur. J. Pharm. Biopharm.* **2007**, *66*, 303–317. *Poly(ethylene oxide)–poly(propylene oxide) block copolymer micelles as drug delivery agents: Improved hydrosolubility, stability and bioavailability of drugs.*

[45]   R. Vakil, G. S. Kwon, *Langmuir* **2006**, *22*, 9723–9729. *Poly(ethylene glycol)-b-poly(e-caprolactone) and PEG-phospholipid form stable mixed micelles in aqueous media.*

[46]   K. Osada, R. J. Christie, K. Kataoka, *J. R. Soc. Interface* **2009**, *6*, S325–S339. *Polymeric micelles from poly(ethylene glycol)–poly(amino acid) block copolymer for drug and gene delivery.*

[47]   S. K. Agrawal, N. Sanabria-DeLong, J. M. Coburn, G. N. Tew, S. R. Bhatia, *J. Control. Release* **2006**, *112*, 64–71. *Novel drug release profiles from micellar solutions of PLA–PEO–PLA triblock copolymers.*

[48]   X. Yu, Y. Zhang, Ch. Chen, Q. Yao, M. Li, *Biochim. Biophys. Acta* **2010**, *1805*, 97–104. *Targeted drug delivery in pancreatic cancer.*

[49]   A. C Misra, S. Bhaskar, N. Clay, J. Lahann, *Adv. Mater.* **2012**, *24*, 3850–3856. *Multicompartmental particles for combined imaging and siRNA delivery.*

[50]   J. Xu, Y. Jiao, X. Shao, Ch. Zhou, *Mater. Lett.* **2011**, *65*, 2800–2803. *Controlled dual release of hydrophobic and hydrophilic drugs from electrospun poly(l-lactic acid) fiber mats loaded with chitosan microspheres.*

[51]   S. Fusco, A. Borzacchiello, P. A. Netti, *J. Bioact. Compat. Polym.* **2006**, *21*, 149–164. *Perspectives on: PEO-PPO-PEO triblock copolymers and their biomedical applications.*

[52]   A. B. Saim, Y. Cao, Y. Weng, Ch.-N. Chang, M. A. Vacanti, Ch. A. Vacanti, R. D. Eavey, *Laryngoscope* **2000**, *110*, 1694–1697. *Engineering autogenous cartilage in the shape of a helix using an injectable hydrogel scaffold.*

[53]   N. Das, T. Bera, A. Mukherjee, *Int. J. Pharm. Bio. Sci.* **2012**, *3*, 586–597. *Biomaterial hydrogels for different biomedical applications.*

[54]   J. L. West, J. A. Hubbell, *React. Polym.* **1995**, *25*, 139–147. *Photopolymerized hydrogel materials for drug delivery applications.*

[55]   B. D. Polizzotti, B. D. Fairbanks, K. S. Anseth, *Biomacromolecules* **2008**, *9*, 1084–1087. *Three-dimensional biochemical patterning of Click-based composite hydrogels via thiol-ene photopolymerization.*

[56]   M. Malkoch, R. Vestberg, N. Gupta, L. Mespouille, P. Dubois, A. Mason A, J. L. Hedrick, Q. Liao, C. W. Frank, K. Kingsbury, C. J. Hawker, *Chem. Commun.* **2006**, 2774–2776. *Synthesis of well-defined hydrogel networks using Click chemistry.*

[57]   J. Zhu, *Biomaterials* **2010**, *31*, 4639–4656. *Bioactive modification of poly(ethylene glycol) hydrogels for tissue engineering.*

[58]   H. Lee, B. G. Choi, H. J. Moon, J. Choi, K. Park, B. Jeong, D. K. Han, *Macromol. Res.* **2012**, *20*, 106–111. *Chondrocyte 3D-culture in RGD-modified crosslinked hydrogel with temperature-controllable modulus.*

[59]   S. Bhaskar, J. Hitt, S.-W. L. Chang, J. Lahann, *Angew. Chem. Int. Ed.* **2009**, *48*, 4589–4593. *Multicompartmental microcylinders.*

[60]   S. Saha, D. Copic, S. Bhaskar, N. Clay, A. Donini, A. J. Hart, J. Lahann, *Angew. Chem. Int. Ed.* **2012**, *51*, 660–665. *Chemically controlled bending of compositionally anisotropic microcylinders.*

[61] M. S. Thompson, T. P. Vadala, M. L. Vadala, Y. Lin, J. S. Riffle, *Polymer* **2008**, *49*, 345–373. *Synthesis and applications of heterobifunctional poly(ethylene oxide) oligomers.*

[62] J. Zhang, Y.-J. Zhao, Zh.-G. Su, G.-H. Ma. *J. Appl. Polym. Sci.* **2007**, *105*, 3780–3786. *Synthesis of monomethoxy poly(ethylene glycol) without diol poly(ethylene glycol).*

[63] T. Ishii, M. Yamada, T. Hirase, Y. Nagasaki, *Polym. J.* **2005**, *37*, 221–228. *New Synthesis of heterobifunctional poly(ethylene glycol) possessing a pyridyl disulfide at one end and a carboxylic acid at the other end.*

[64] Z. Li, Y. Chau, *Bioconjugate Chem.* **2011**, *22*, 518–522. *Synthesis of X(Y)-(EO)$_n$-OCH$_3$ type heterobifunctional and X(Y)-(EO)$_n$-Z type heterotrifunctional poly(ethylene glycol)s.*

[65] S. Herrwerth, T. Rosendahl, C. Feng, J. Fick, W. Eck, M. Himmelhaus, R. Dahint, M. Grunze, *Langmuir* **2003**, *19*, 1880–1887. *Covalent coupling of antibodies to self-assembled monolayers of carboxy-functionalized poly(ethylene glycol): Protein resistance and specific binding of biomolecules.*

[66] C. E. Hoyle, Ch. N. Bowman, *Angew. Chem. Int. Ed.* **2010**, *49*, 1540–1573. *Thiol-ene click chemistry.*

[67] F. Zeng, Ch. Allen, *Macromolecules* **2006**, *39*, 6391–6398. *Synthesis of Carboxy-Functionalized heterobifunctional poly(ethylene glycol) by a thiol-anionic polymerization method.*

[68] Sh. Zhang, J. Du, R. Sun, X. Li, D. Yang, Sh. Zhang, Ch. Xiong, Y. Peng, *React. Funct. Polym.* **2003**, *56*, 17–25. *Synthesis of heterobifunctional poly(ethylene glycol) with a primary amino group at one end and a carboxylate group at the other end.*

[69] Y. Nagasaki, M. Iijima, M. Kato, K. Kataoka, *Bioconjugate Chem.* **1995**, *6*, 702–704. *Primary amino-terminal heterobifunctional poly(ethylene oxide). Facile synthesis of poly(ethylene oxide) with a primary amino group at one end and a hydroxyl group at the other end.*

[70]   K. Sui, L. Gu, *J. App. Polym. Sci.* **2003**, *89*, 1753–1759. *Preparation and characterization of amphiphilic block copolymer of polyacrylonitrile-block-poly(ethylene oxide)).*

[71]   M. Yokoyama, T. Okano, Y. Sakurai, *Bioconjugate Chem.* **1992**, *3*, 275–276. *Synthesis of poly(ethylene oxide) with heterobifunctional reactive groups at its terminals by an anionic initiator.*

[72]   Y. J. Kim, Y. Nagasaki, K. Kataoka, M. Kato, M. Yokoyama, T. Okano, Y. Sakurai, *Polym. Bull.* **1994**, *33*, 1–6. *Heterobifunctional poly(ethylene oxide). One pot synthesis of poly(ethylene oxide) with a primary amino group at one end and a hydroxyl group at the other end.*

[73]   Y. Nagasaki, T. Kutsuna, M. Iijima, M. Kato, K. Kataoka, *Bioconjugate Chem.* **1995**, *6*, 231–233. *Formyl-ended heterobifunctional poly(ethylene oxide): Synthesis of poly(ethylene oxide) with a formyl group at one end and a hydroxyl group at the other end.*

[74]   M. Oishi, Sh. Sasaki, Y. Nagasaki, K. Kataoka, *Biomacromolecules* **2003**, *4*, 1426–1432. *pH-Responsive oligodeoxynucleotide (ODN)-poly(ethylene glycol) conjugate through acid-labile α-thiopropionate linkage: preparation and polyion complex micelle formation.*

[75]   Y. Akiyama, Y. Nagasaki, K. Kataoka, *Bioconjugate Chem.* **2004**, *15*, 424–427. *Synthesis of heterotelechelic poly(ethylene glycol) derivatives having alpha-benzaldehyde and omega-pyridyl disulfide groups by ring opening polymerization of ethylene oxide using 4-(diethoxymethyl)benzyl alkoxide as a novel initiator.*

[76]   Sh. Hiki, K. Kataoka, *Bioconjugate Chem.* **2010**, *21*, 248–254. *Versatile and selective synthesis of "click chemistry" compatible heterobifunctional poly(ethylene glycol)s possessing azide and alkyne functionalities.*

[77]   C. Mangold, F. Wurm, B. Obermeier, H. Frey, *Macromol. Rapid Commun.* **2010**, *31*, 258–264. *Hetero-multifunctional poly(ethylene glycol) copolymers with multiple hydroxyl groups and a single terminal functionality.*

[78] A. Aqil, H. Serwas, J. L. Delplancke, R. Jerome, C. Jerome, L. Canet, *Ultrason. Sonochem.* **2008**, *15*, 1055–1061. *Preparation of stable suspensions of gold nanoparticles in water by sonoelectrochemistry.*

[79] J. Raynaud, Ch. Absalon, Y. Gnanou, D. Taton, J. Am. Chem. Soc. **2009**, *131*, 3201–3209. *N-Heterocyclic carbene-induced zwitterionic ring-opening polymerization of ethylene oxide and direct synthesis of α,ω-difunctionalized poly(ethylene oxide)s and poly(ethylene oxide)-b-poly(ε-caprolactone) block copolymers.*

[80] J. Raynaud, Ch. Absalon, Y. Gnanou, D. Taton, *Macromolecules* **2010**, *43*, 2814–2823. *N-Heterocyclic carbene-organocatalyzed ring-opening polymerization of ethylene oxide in the presence of alcohols or trimethylsilyl nucleophiles as chain moderators for the synthesis of α,ω-heterodifunctionalized poly(ethylene oxide)s.*

[81] N. H. Völcker, D. Klee, M. Hanna, H. Höcker, J. J. Bou, A. M. de Ilarduya, S. Munoz-Guerra, *Macromol. Chem. Phys.* **1999**, *200*, 1363–1373. *Synthesis of heterotelechelic poly(ethylene glycol)s and their characterization by MALDI-TOF-MS.*

[82] G. Lapienis, S. Penczek, *J. Bioact. Compat. Pol.* **2001**, *16*, 206–220. *Preparation of monomethyl ethers of poly(ethylene glycol)s free of the poly(ethylene glycol).*

[83] R. Mahou, Ch. Wandrey, *Polymers* **2012**, *4*, 561–589. *Versatile route to synthesize heterobifunctional poly(ethylene glycol) of variable functionality for subsequent pegylation.*

[84] S. Svenson, *Eur. J. Pharm. Biopharm.* **2009**, *71*, 445–462. *Dendrimers as versatile platform in drug delivery applications.*

[85] M. Zacchigna, F. Cateni, S. Drioli, G. M. Bonora, *Polymers* **2011**, *3*, 1076–1090. *Multimeric, multifunctional derivatives of poly(ethylene glycol).*

[86] S. Driolia, G. M. Bonora, M. Ballico, *Open. Org. Chem. J.* **2008**, *2*, 17–25. *New Syntheses of branched, multifunctional high-molecular weight poly(ethylene glycol)s or (multiPEG)s.*

[87] G. Pasut, S. Scaramuzza, O. Schiavon, R. Mendichi, F. M. Veronese, *J. Bioact. Compat. Pol.* **2005**, *20*, 213–230. *PEG-epirubicin conjugates with high drug loading.*

[88]   C. J. Hawker, F. Chu, P. J. Pomery, D. J. T. Hill, *Macromolecules* **1996**, *29*, 3831–3838. *Hyperbranched poly(ethylene glycol)s: A new class of ion-conducting materials.*

[89]   X.-Sh. Feng, D. Taton, E. L. Ch., Y. Gnanou, *J. Am. Chem. Soc.* **2005**, *127*, 10956–10966. *Toward an easy access to dendrimer-like poly(ethylene oxide)s.*

[90]   M. Zacchignaa, G. Di Lucaa, F. Catenia, V. Mauricha, M. Ballico, G. M. Bonorab, S. Drioli, *Eur. J. Pharm. Sci.* **2007**, *30*, 343–350. *New multiPEG-conjugated theophylline derivatives: Synthesis and pharmacological evaluations.*

[91]   W. Kuran, *Prog. Polym. Sci.* **1998**, *23*, 919–992. *Coordination polymerization of heterocyclic and heterounsaturated monomers.*

[92]   C. C. Price, Y. Atarashi, R. Yamamoto, *J. Polym. Sci., Part A: Polym. Chem.* **1969**, *7*, 569–574. *Polymerization and copolymerization of some epoxides by potassium tert-butoxide in DMSO.*

[93]   G.-E. Yu, A. J Masters, F. Heatley, C. Booth, T. G. Blease, *Makromol. Chem.* **1986**, *181*, 745–752. *Anionic polymerisation of propylene oxide. Investigation of double-bond and head-to-head content by NMR spectroscopy.*

[94]   J. Allgaier, S. Willbold, T. Chang, *Macromolecules* **2007**, *40*, 518–525. *Synthesis of hydrophobic poly(alkylene oxide)s and amphiphilic poly(alkylene oxide) block copolymers.*

[95]   M. I. Malik, B. Trathnigg, C. O. Kappe, *Eur. Polym. J.* **2009**, *45*, 899–910. *Microwave-assisted polymerization of higher alkylene oxides.*

[96]   C. Mangold, F. Wurm, H. Frey, *Polym. Chem.* **2012**, *3*, 1714–1721. *Functional PEG-based polymers with reactive groups via anionic ROP of tailor-made epoxides.*

[97]   B. F. Lee, M. J. Kade, J. A. Chute, N. Gupta, L. M. Campos, G. H. Fredrickson, E. J. Kramer, N. A. Lynd, C. J. Hawker, *J. Polym. Sci., Part A: Polym. Chem.* **2011**, *49*, 4498–4504. *Poly(allyl glycidyl ether) – a versatile and functional polyether platform.*

[98]   B. Obermeier, Holger Frey, *Bioconjugate Chem.* **2011**, *22*, 436–444. *Poly(ethylene glycol-co-allyl glycidyl ether)s: A PEG-Based modular synthetic platform for multiple bioconjugation.*

[99]   M. Erberich, H. Keul, M Moeller, *Macromolecules* **2007**, *40*, 3070–3079. *Polyglycidols with two orthogonal protective groups: Preparation, selective deprotection, and functionalization.*

[100]   J. N. Hunt , K. E. Feldman , N. A. Lynd , J. Deek , L. M. Campos, J. M. Spruell , B. M. Hernandez , E. J. Kramer, C. J. Hawker, *Adv. Mater.* **2011**, *23*, 2327–2331. *Tunable, high modulus hydrogels driven by ionic coacervation.*

[101]   C. Mangold, C. Dingels, B. Obermeier, H. Frey, F. Wurm, *Macromolecules* **2011**, *44*, 6326–6334. *PEG-based multifunctional polyethers with highly reactive vinyl-ether side chains for click-type functionalization.*

[102]   R. Su, Y. Qin, L. Qiao, J. Li, X. Zhao, P. Wang, X. Wang, F. Wang, *J. Polym. Sci., Part A: Polym. Chem.* **2011**, *49*, 1434–1442. *Synthesis of poly(2-furyloxirane) with high molecular weight and improved regioregularity using macrocyclic ether as a cocatalyst to potassium tert-butoxide.*

[103]   R. M. Thomas, P. C. B. Widger, S. M. Ahmed, R. C. Jeske, W. Hirahata, E. B. Lobkovsky, G. W. Coates, *J. Am. Chem. Soc.* **2010**, *132*, 16520–16525. *Enantioselective epoxide polymerization using a bimetallic cobalt catalyst.*

[104]   Y. Hu, L. Qiao, Y. Qin, X. Zhao, X. Chen, X. Wang, F. Wang, *Macromolecules* **2009**, *42*, 9251–9254. *Synthesis and stabilization of novel aliphatic polycarbonate from renewable resource.*

[105]   A. Gandini, *Prog. Polym. Sci.* **2013**, *38*, 1–29. *The furan/maleimide Diels–Alder reaction: A versatile click–unclick tool in macromolecular synthesis.*

[106]   J. Meyer, H. Keul, M. Möller, *Macromolecules* **2011**, *44*, 4082–4091. *Poly(glycidyl amine) and copolymers with glycidol and glycidyl amine repeating units: Synthesis and characterization.*

[107] H.-Q. Xie, J.-Sh. Guo, G.-Q. Yu, J. Zu, *J. Appl. Polym. Sci.* **2001**, *80*, 2446–2454. *Ring-opening polymerization of epichlorohydrin and its copolymerization with other alkylene oxides by quaternary catalyst system.*

[108] S. Carlotti, A. Labbe, V. Rejsek, S. Doutaz, M. Gervais, A. Deffieux, *Macromolecules* **2008**, *41*, 7058–7062. *Living/controlled anionic polymerization and copolymerization of epichlorohydrin with tetraoctylammonium bromide-triisobutylaluminum initiating systems.*

[109] A.-L. Brocas, G. Cendejas, S. Caillol, A. Deffieux, S. Carlotti, *J. Polym. Sci., Part A: Polym. Chem.* **2011**, *49*, 2677–2684. *Controlled synthesis of polyepichlorohydrin with pendant cyclic carbonate functions for isocyanate-free polyurethane networks.*

[110] T. Satoh, H Ishihara, H. Sasaki, H. Kaga, T. Kakuchi, *Macromolecules* **2003**, *36*, 1522–1525. *A novel ladder polymer. Two-step polymerization of oxetanyl oxirane leading to a "fused 15-crown-4 polymer" having a high Li$^+$-binding ability.*

[111] S. M. Ramirez, J. M. Layman, Ph. Bissel, T. E. Long, *Macromolecules* **2009**, *42*, 8010–8012. *Ring-opening polymerization of imidazole epoxides for the synthesis of imidazole-substituted poly(ethylene oxides).*

[112] Ch. Tonhauser, A. Alkan, M. Schömer, C. Dingels, S. Ritz, V. Mailänder, H. Frey, F. R. Wurm, *Macromolecules* **2013**, *46*, 647–655. *Ferrocenyl glycidyl ether: A versatile ferrocene monomer for copolymerization with ethylene oxide to water-soluble, thermoresponsive copolymers.*

[113] H. Keul, M. Möller, *J. Polym. Sci., Part A: Polym. Chem.* **2009**, *47*, 3209–3231. *Synthesis and degradation of biomedical materials based on linear and star shaped polyglycidols.*

[114] M. Schömer, Ch. Schüll, H. Frey, *J. Polym. Sci., Part A: Polym. Chem.* **2013**, *51*, 995–1019. *Hyperbranched aliphatic polyether polyols.*

[115] H. Misaka, E. Tamura, K. Makiguchi, K. Kamoshida, R. Sakai, T. Satoh, T. Kakuchi, *J. Polym. Sci., Part A: Polym. Chem.* **2012**, *50*, 1941–1952. *Synthesis of end-*

functionalized polyethers by phosphazene base-catalyzed ring-opening polymerization of 1,2-butylene oxide and glycidyl ether.

[116]   J. Geschwind, H. Frey, *Macromolecules* **2013**, *46*, 3280–3287. *Poly(1,2-glycerol carbonate): A fundamental polymer structure synthesized from CO_2 and glycidyl ethers.*

[117]   C. Mangold, F. Wurm, B. Obermeier, H. Frey, *Macromolecules* **2010**, *43*, 8511–8518. *"Functional poly(ethylene glycol)": PEG-based random copolymerswith 1,2-diol side chains and terminal amino functionality.*

[118]   P. Goutte, M. Sepulchre, N. Spassky, *Makromol. Chem., Macromol. Symp.*, **1986**, *6*, 225-235. *Synthesis of polyoxiranes and polythiiranes bearing a potential aldehyde group in the side chain.*

[119]   B. Obermeier, F. Wurm, H. Frey, *Macromolecules* **2010**, *43*, 2244–2251. *Amino functional poly(ethylene glycol) copolymers via protected amino glycidol.*

[120]   V. S. Reuss, B. Obermeier, C. Dingels, H. Frey, *Macromolecules* **2012**, *45*, 4581–4589. *N,N-Diallylglycidylamine: A key monomer for amino-functional poly(ethylene glycol) architectures.*

[121]   M. A. Gauthier, M. I. Gibson, H.-A. Klok, *Angew. Chem. Int. Ed.* **2009**, *48*, 48–58. *Synthesis of functional polymers by post-polymerization modification.*

[122]   H. Nandivada, X. Jiang, J. Lahann, *Adv. Mater.* **2007**, *19*, 2197–2208. *Click chemistry: Versatility and control in the hands of materials scientists.*

[123]   Y. Koyama, M. Umehara, A. Mizuno, M. Itaba, *Bioconjugate Chem.* **1996**, *7*, 298–301. *Synthesis of novel poly(ethylene glycol) derivatives having pendant amino groups and aggregating behavior of its mixture with fatty acid in water.*

[124]   Ch. Yoshihara, Ch.-Y. Shew, T. Ito, Y. Koyama, *Biophys. J.* **2010**, *98*, 1257–1266. *Loosening of DNA/polycation complexes by synthetic polyampholyte to improve the transcription efficiency: effect of charge balance in the polyampholyte.*

[125]   M. Hruby, C. Konak, K. Ulbrich, *J. Control. Release* **2005**, *103*, 137–148. *Polymeric micellar pH-sensitive drug delivery system for doxorubicin.*

[126]   Zh. Hu, X. Fan, J. Wang, Sh. Zhao, X. Han, *J. Polym. Res.* **2010**, *17*, 815–820. *A facile synthesis of hydrophilic glycopolymers with the ether linkages throughout the main chains.*

[127]   J. A. Reina, V. Cadiz, A. Mantecon, A. Serra, *Angew. Chem. Int. Ed.* **1993**, *209*, 95–109. *Chemical modification of poly(epichlorohydrin) with unsaturated potassium carboxylates.*

[128]   M. Perez, J. A. Reina, A. Serra, J. C. Ronda, *Acta Polymer.* **1998**, *49*, 312– 318. *Chemical modification of poly(epichlorohydrin) with phenolate. Studies of the side reactions.*

[129]   F. Bekkar, M. Belbachir, *Chin. J. Chem.* **2009**, *27*, 1174—1178. *Chemical modification of poly(epichlorohydrin) using montmorillonite clay.*

[130]   Zh. Li, P. Li, J. Huang, *Polymer* **2006**, *47*, 5791-5798. *Synthesis and characterization of amphiphilic graft copolymer poly(ethylene oxide)-graft-poly(methyl acrylate).*

[131]   Zh. Li, Y. Chau, *Bioconjugate Chem.* **2009**, *20,* 780–789. *Synthesis of linear polyether polyol derivatives as new materials for bioconjugation.*

[132]   G. Odian. Principles of Polymerization. Fourth Edition; John Wiley & Sons, Inc.: Hoboken, New Jersey, 2010; Chapter 7, pp. 544–618.

[133]   F. M. Veronese, *Biomaterials* **2001**, *22*, 405–417. *Peptide and protein PEGylation: A review of problems and solutions.*

[134]   K. S. Kazanskii, A. A. Solovyanov, S. G. Entelis, *Eur. Polym. J.* **1971**, *7*, 1421–1433. *Polymerization of ethylene oxide by alkali metal-naphthalene complexes in tetrahydrofuran.*

[135]   J. Furukawat, T. Saegusa, T. Tsuruta, G. Kakogawa, *Makromol. Chem.* **1959**, *31*, 25–39. *New catalyst for the polymerization of alkylene oxides.*

[136] S. Boileau, N. Illy, *Prog. Polym. Sci.* **2011**, *36*, 1132–1151. *Activation in anionic polymerization: Why phosphazene bases are very exciting promoters.*

[137] V. Rejsek, Ph. Desbois, A. Deffieux, S. Carlotti, *Polymer* **2010**, *51*, 5674–5679. *Polymerization of ethylene oxide initiated by lithium derivatives via the monomer-activated approach: Application to the direct synthesis of PS-b-PEO and PI-b-PEO diblock copolymers.*

[138] M. D. Shalati, C. G. Overberger, *J. Polym. Sci.: Polym. Chem. Ed.* **1983**, *21*, 3425–3442. *Grafting of living poly(ethylene oxide) onto polystyrene via aromatic nucleophilic displacement of activated nitro groups.*

[139] A. Stolarzewicz, D. Neugebauer, *Macromol. Chem. Phys.* **1998**, *199*, 175–177. *Influence of the kind of crown ether on the heterogeneous polymerization of propylene oxide in the presence of potassium hydride.*

[140] A. Dworak, I. Panchev, B. Trzebicka, W. Walach, *Macromol. Symp.* **2000**, *153*, 233–242. *Hydrophilic and amphiphilic copolymers of 2,3-epoxypropanol-1.*

[141] Y.-H. Huang, Z.-M. Li, H. Morawetz, *J. Polym. Sci.: Polym. Chem. Ed.* **1985**, *23*, 795–799. *The kinetics of the attachment of polymer chains to reactive latex particles and the resulting latex stabilization.*

[142] J. Huang, H. Wang, X. Tian, *J. Polym. Sci., Part A: Polym. Chem.* **1996**, *34*, 1933–1940. *Preparation of PEO with amine and sulfadiazine end groups by anion ring-opening polymerization of ethylene oxide.*

[143] M. Hruby, C Konak, K. Ulbrich, *J. Appl. Polym. Sci.* **2005**, *95*, 201–211. *Poly(allyl glycidyl ether)-block-poly(ethylene oxide): A novel promising polymeric intermediate for the preparation of micellar drug delivery systems.*

[144] M. Hans, H. Keul, M. Möller, *Polymer* **2009**, *50*, 1103–1108. *Chain transfer reactions limit the molecular weight of polyglycidol prepared via alkali metal based initiating systems.*

[145]   I. C. Stewart, C. C. Lee, R. G. Bergman, F. D. Toste, *J. Am. Chem. Soc.* **2005**, *127*,
        17616–17617. *Living ring-opening polymerization of N-sulfonylaziridines: Synthesis
        of high molecular weight linear polyamines.*

[146]   Y. Yagci, K. Ito, *Macromol. Symp.* **2005**, *226*, 87–96. *Macromolecular architecture
        based on anionically prepared poly(ethylene oxide) macromonomers.*

[147]   R. K. Iha, K. L. Wooley, A. M. Nystroem, D. J. Burke, M. J. Kade, C. J. Hawker
        Chem. Rev. **2009**, *109*, 5620–5686. *Applications of orthogonal "click" chemistries in
        the synthesis of functional soft materials.*

[148]   S. D. Patterson, V. Katta, *Anal. Chem.* **1994**, *66*, 3121–3132. *Prompt fragmentation of
        disulfide-linked peptides during matrix-assisted laser desorption ionization mass
        spectrometry.*

[149]   A. J. Birch, *J. Chem. Soc.*, **1945**, 809–813. *Reduction by dissolving metals. Part II.*

[150]   T. Hanazawa, K. Sasaki, Y. Takayama, F. Sato, *J. Org. Chem.* **2003**, *68*, 4980–4983.
        *Efficient and practical method for synthesizing optically active indan-2-ols by the
        Ti(O-i-Pr)₄/2 i-PrMgCl-mediated metalative Reppe reaction.*

[151]   D. M. Troast, J. Yuan, John A. Porco Jr., *Adv. Synth. Catal.* **2008**, *350*, 1701 – 1711.
        Studies toward the synthesis of (–)-zampanolide: Preparation of the macrocyclic core.

[152]   T. Saito, M. Morimoto, Ch. Akiyama, T. Matsumoto, K. Suzuki, *J. Am. Chem. Soc.*
        1995, *117*, 10757–10758. *Stereocontrolled convergent total synthesis of (±)-
        furaquinocin D.*

[153]   A. G. Myers, Ph. M. Harrington, E. Y. Kuo, *J. Am. Chem. Soc.* **1991**, *113*, 694–695.
        Enantioselective synthesis of the epoxy diyne core of neocarzinostatin chromophore.

[154]   Y. Furukawa, Y. Miki, *Eur. Pat. Appl.* **2004**, EP 1403267 A1, 20040331. *Process for
        preparing glycidylphthalimide.*

[155]   P. Stangier, O. Hindsgaul, *Synlett* **1996**, *2*, 179–181. *Solid-phase transimidation for
        the removal of N-phthalimido- and N-tetrachlorophthalimido protecting groups on
        carbohydrates.*

[156] A. O. Fitton, J. Hill, D. E. Jane, R. Millar, *Synthesis* **1987**, 1140–1142. *Synthesis of simple oxetanes carrying reactive 2-substituents.*

[157] S. M. Sirard, H. J. Castellanos, H. S. Hwang, K.-T. Lim, K. P. Johnston, *Ind. Eng. Chem. Res.*, **2004**, *43*, 525–534. *Steric stabilization of silica colloids in supercritical carbon dioxide.*

[158] A. Pfenninger, *Synthesis* **1986**, *2*, 89–116. *Asymmetric epoxidation of allylic alcohols: The Sharpless epoxidation.*

[159] F. Charmantray, P. Dellis, V. Hélaine, S. Samreth, L. Hecquet. *Eur. J. Org. Chem.* **2006**, *24*, 5526–5532. *Chemoenzymatic synthesis of 5-thio-D-xylopyranose.*

[160] C. Tonhauser, H. Frey, *Macromol. Rapid Commun.* **2010**, *31*, 1938–1947. *A road less traveled to functional polymers: Epoxide termination in living carbanionic polymer synthesis.*

[161] F. Heatley, G. Yu, C. Booth, T. G. Blease, *Eur. Polym. J.* **1991**, *27*, 573–579. *Determination of reactivity ratios for the anionic copolymerization of ethylene oxide and propylene oxide in bulk.*

[162] B. F. Lee, M. Wolff, K. T. Delaney, J. K. Sprafke, F. A. Leibfarth, C. J. Hawker, N. A. Lynd, *Macromolecules* **2012**, *45*, 3722–3731. *Reactivity ratios and mechanistic insight for anionic ring-opening copolymerization of epoxides.*

[163] A. Stolarzewicz, *Makromol. Chem.* **1986**, *181*, 745–752. *A new chain transfer reaction in the anionic polymerization of 2,3-epoxypropyl phenyl ether and other oxiranes.*

[164] Y.-J. Huang, G.-R. Qi, Y.-H. Wang, *J. Polym. Sci., Part A: Polym. Chem.* **2002**, *40*, 1142–1150. *Controlled ring-opening polymerization of propylene oxide catalyzed by double metal-cyanide complex.*

[165] I. Kim, J.-T. Ahn, Ch. S. Ha, Ch. S. Yang, I. Park, *Polymer* **2003**, *44*, 3417–3428. *Polymerization of propylene oxide by using double metal cyanide catalysts and the application to polyurethane elastomer.*

[166] A. Dworak, B. Trzebicka, W. Walach, A. Utrata, C. Tsvetanov, *Macromol. Symp.* **2004**, *210*, 419–426. *Novel reactive thermosensitive polyethers – control of transition point.*

[167] S. Rangelov, B. Trzebicka, M. Jamroz-Piegza, A. Dworak, *J. Phys. Chem. B* **2007**, *111*, 11127–11133. *Hydrodynamic behavior of high molar mass linear polyglycidol in dilute aqueous solution.*

[168] M. Gervais, A.-L. Brocas, G. Cendejas, A. Deffieux, S. Carlotti, *Macromolecules* **2010**, *43*, 1778–1784. *Synthesis of linear high molar mass glycidol-based polymers by monomer-activated anionic polymerization.*

[169] F. Guibe, *Tetrahedron* **1997**, *53*, 13509–13556. *Allylic protecting groups and their use in a complex environment. Part I: Allylic protection of alcohols.*

[170] H. Nandivada, A. M. Ross, Joerg Lahann, *Prog. Polym. Sci.* **2010**, *35*, 141–154. *Stimuli-responsive monolayers for biotechnology.*

[171] E. Engel, A. Michiardi, M. Navarro, D. Lacroix, J. A. Planell, *Trends Biotechnol.* **2008**, *26*, 39–47. *Nanotechnology in regenerative medicine: The materials side.*

[172] O. Zinger, G. Zhao, Z. Schwartz, J. Simpson, M. Wieland, D. Landolt, B. Boyan, *Biomaterials* **2005**, 26, 1837–1847. *Differential regulation of osteoblasts by substrate microstructural features.*

[173] D. Lehnert, B. Wehrle-Haller, Ch. David, U. Weiland, Ch. Ballestrem, B. A. Imhof, M. Bastmeyer, *J. Cell Sci.* **2004**, *117*, 41–52. *Cell behaviour on micropatterned substrata: Limits of extracellular matrix geometry for spreading and adhesion.*

[174] V. A. Schulte, Y. Hu, M. Diez, D. Bünger, M. Möller, M. C. Lensen. *Biomaterials* **2010**, *31*, 8583–8595. *A hydrophobic perfluoropolyether elastomer as a patternable biomaterial for cell culture and tissue engineering.*

[175] T. T. Truong, R. Lin, S. Jeon, H. H. Lee, J. Maria, A. Gaur, F. Hua, I. Meinel, J. A. Rogers, *Langmuir* **2007**, *23*, 2898–2905. *Soft lithography using acryloxy perfluoropolyether composite stamps.*

[176] C. A. Scotchford, M. Ball, M. Winkelmann, J. Vörös, C. Csucs, D. M. Brunette, G. Danuser, M. Textor, *Biomaterials* **2003**, *24,* 1147–1158. *Chemically patterned, metal-oxide-based surfaces produced by photolithographic techniques for studying protein- and cell-interactions. II: Protein adsorption and early cell interactions.*

[177] M. S. Hahn, L. J. Taite, J. J. Moon, M. C. Rowland, K. A. Ruffino, J. L. West, *Biomaterials* **2006**, *27,* 2519–2524. *Photolithographic patterning of polyethylene glycol hydrogels.*

[178] M. Kim, J.-Ch. Choi, H.-R. Jung, J. S. Katz, M.-G. Kim, J. Doh, *Langmuir* **2010**, *26,* 12112–12118. *Addressable micropatterning of multiple proteins and cells by microscope projection photolithography based on a protein friendly photoresist.*

[179] H. Y. Chen, M. Hirtz, X. Deng, T. Laue, H. Fuchs, J. Lahann, *J. Am. Chem. Soc.* **2010**, *132,* 18023–18025. *Substrate-independent dip-pen nanolithography based on reactive coatings.*

[180] P. M. Mendes, Ch. L. Yeung, J. A. Preece, *Nanoscale Res. Lett.* **2007**, *2,* 373–384. *Bio-nanopatterning of surfaces.*

[181] R. D. Piner, J. Zhu, F. Xu, S. Hong, Ch. A. Mirkin, *Science* **1999**, *283,* 661–663. *"Dip-pen" nanolithography.*

[182] A. D. Moorhouse, J. E. Moses, Chem. Med. Chem. **2008**, *3,* 715–723. *Click chemistry and medicinal chemistry: A case of "cyclo-addiction".*

[183] J. C. Jewetta, C. R. Bertozzi, *Chem. Soc. Rev.* **2010**, *39,* 1272–1279. *Cu-free click cycloaddition reactions in chemical biology.*

[184] S. G. Gouin, J. Kovensky, *Synlett* **2009**, *9,* 1409 1412. *A procedure for fast and regioselective copper-free click chemistry at room temperature with p-toluenesulfonyl alkyne.*

[185] M. Clark, P. Kiser, *Polym. Int.* **2009**, *58,* 1190–1195. *In situ crosslinked hydrogels formed using Cu(I)-free Huisgen cycloaddition reaction.*

[186]   J. Lahann, H. Höcker, R. Langer, *Angew. Chem. Int. Ed.* **2001**, *40*, 726–728. *Synthesis of amino[2.2]paracyclophanes-beneficial monomers for bioactive coating of medical implant materials.*

[187]   A.S. Hoffman, *Adv. Drug Deliv. Rev.*, **2002**, *43*, 3–12. *Hydrogels for biomedical applications.*

[188]   N. A. Peppas, J. Z. Hilt, A. Khademhosseini, R. Langer. *Adv. Mater.* **2006**, *18*, 1345–1360. *Hydrogels in biology and medicine: From molecular principles to bionanotechnology.*

[189]   P P. Kalshetti, V. B. Rajendra, D. N. Dixit, P. P. Parekh, *Int. J. Pharm. Pharm. Sci.* **2012**, *4*, 1–7. *Hydrogels as a drug delivery system and applications: A review.*

[190]   R. Censi, P. Di Martino, T. Vermonden, W. E. Hennink, *J. Control. Release* **2012**, *161*, 680–692. *Hydrogels for protein delivery in tissue engineering.*

[191]   S. R. Van Tomme, G. Storm, W. E. Hennink, *Int. J. Pharm.* **2008**, *355*, 1–18. *In situ gelling hydrogels for pharmaceutical and biomedical applications.*

[192]   J. Maia, L. Ferreira, R. Carvalho, M. A. Ramos, M. H. Gil, *Polymer* **2005**, *46*, 9604–9614. *Synthesis and characterization of new injectable and degradable dextran-based hydrogels.*

[193]   H. Yu, J. Li, D. Wu, Zh. Qiu, Y. Zhang, *Chem. Soc. Rev.* **2010**, *39*, 464–473. *Chemistry and biological applications of photo-labile organic molecules.*

[194]   A. P. Pelliccioli, J. Wirz, *Photochem. Photobiol. Sci.* **2002**, *1*, 441–458. *Photoremovable protecting groups: Reaction mechanisms and applications.*

[195]   Ch. G. Bochet, *J. Chem. Soc., Perkin Trans. 1*, **2002**, 125–142. *Photolabile protecting groups and linkers.*

[196]   P. Wang, H. Hu, Y Wang. *Org. Lett.* **2007**, *9*, 1533–1535. *Novel photolabile protecting group for carbonyl compounds.*

[197]   N. C. Shinde, N. J. Keskar, P. D. Argade, *Res. J. Pharm., Biol. Chem. Sci.* **2012**, *3*, 922–929. *Nanoparticles: Advances in drug delivery systems.*

[198] V. Wood, M. J. Panzer, J.-M. Caruge, J. E. Halpert, M. G. Bawendi, V. Bulovic, *Nano Lett.* **2010**, *10*, 24–29. *Air-stable operation of transparent, colloidal quantum dot based LEDs with a unipolar device architecture.*

[199] S. Lal, S. Link, N. J. Halas, *Nature Photon.* **2007**, *1*, 641–648. *Nano-optics from sensing to waveguiding.*

[200] A. Z. Moshfegh, *J. Phys. D: Appl. Phys.* **2009**, *42*, 233001/1–233001/30. *Nanoparticle catalysts.*

[201] M. B. Oliveira, J. F. Mano, *Biotechnol. Prog.* **2011**, *27*, 897–912. *Polymer-based microparticles in tissue engineering and regenerative medicine.*

[202] K. B. Cederquist, S. L. Dean, C. D. Keating, *Wiley Interdiscip. Rev.: Nanomed. Nanobiotechnol.* **2010**, *2*, 578–600. *Encoded anisotropic particles for multiplexed bioanalysis.*

[203] S. Mitragotri, J. Lahann, *Nature Mater.* **2009**, *8*, 15–23. *Physical approaches to biomaterial design.*

[204] K. J. Lee, J. Yoon, J. Lahann *Curr. Opin. Colloid Interface Sci.* **2011**, *16*, 195–202. *Recent advances with anisotropic particles.*

[205] A. B. Pawar, I. Kretzschmar, Macromol. *Rapid Commun.* **2010**, *31*, 150–168. *Fabrication, assembly, and application of patchy particles.*

[206] J. Lahann, *Small* **2011**, *7*, 1149–1156. *Recent progress in nano-biotechnology: Compartmentalized micro- and nanoparticles via electrohydrodynamic co-jetting.*

[207] S. Bhaskar, K. M. Pollock, M. Yoshida, J. Lahann, *Small* **2010**, *6*, 404–411. *Towards designer microparticles: Simultaneous control of anisotropy, shape, and size.*

[208] K.-H. Roh, D. C. Martin, J. Lahann, *Nature Mater.* **2005**, *4*, 759–763. *Biphasic Janus particles with nanoscale anisotropy.*

[209] B. G Davis, M. A. Robinson, *Curr. Opin. Drug Discov. Devel.* **2002**, *5*, 279–288. *Drug delivery systems based on sugar-macromolecule conjugates.*

[210] L. Zhou, R. Cheng, H. Tao, Sh. Ma, W. Guo, F. Meng, H. Liu, Zh. Liu, Zh. Zhong, *Biomacromolecules* **2011**, *12*, 1460–1467. *Endosomal pH-activatable poly(ethylene oxide)-graft-doxorubicin prodrugs: Synthesis, drug release, and biodistribution in tumor-bearing mice.*

[211] A. Sun, J. Lahann, *Soft Matter.* **2009**, *5*, 1555–1561. *Dynamically switchable biointerfaces.*

[212] H. Nandivada, A. M. Ross, J. Lahann, *Prog. Polym. Sci.* **2010**, *35*, 141–154. *Stimuli-responsive monolayers for biotechnology.*

[213] D. Roy, J. N. Cambre, B. S. Sumerlin, *Prog. Polym. Sci.* **2010**, *35*, 278–301. *Future perspectives and recent advances in stimuli-responsive materials.*

[214] J. Kim, R. C. Hayward, *Trends Biotechnol.* **2012**, *30*, 426–439. *Mimicking dynamic in vivo environments with stimuli-responsive materials for cell culture.*

[215] Y. L. Colson, M. W. Grinstaff, *Adv. Mater.* **2012**, *24*, 3878–3886. *Biologically responsive polymeric nanoparticles for drug delivery.*

[216] M. Behl, M. Y. Razzaq, A. Lendlein, *Adv. Mater.* **2010**, *22*, 3388–3410. *Multifunctional shape-memory polymers.*

[217] W. Gao, J. M. Chan, O. C. Farokhzad, *Mol. Pharm.* **2010**, *7*, 1913–1920. *pH-Responsive nanoparticles for drug delivery.*

[218] C. D. Vo, G. Kilcher, N. Tirelli, *Macromol. Rapid Commun.* **2009**, *30*, 299–315. *Polymers and sulfur: What are organic polysulfides good for? Preparative strategies and biological applications.*

[219] N. Yui, T. Okano, Y. Sakurai, *J. Control. Release* **1992**, *22*, 105–116. *Inflammation responsive degradation of crosslinked hyaluronic acid gels.*

[220] K. E. Broaders, S. Grandhe, J. M. J. Frechet, *J. Am. Chem. Soc.* **2011**, *133*, 756–758. *A biocompatible oxidation-triggered carrier polymer with potential in therapeutics.*

[221] A. Rehor, J. A. Hubbell, N. Tirelli, *Langmuir* **2005**, *21*, 411–417. *Oxidation-sensitive polymeric nanoparticles.*

[222]   S. T. Reddy, A. Rehor, H. G. Schmoekel, J. A. Hubbell, M. A. Swartz, *J. Control. Release* **2006**, *112*, 26–34. *In vivo targeting of dendritic cells in lymph nodes with poly(propylene sulfide) nanoparticles.*

[223]   S. T. Reddy, A. J. van der Vlies, E. Simeoni, V. Angeli, G. J. Randolph, C. P. O'Neil, L. K. Lee, M. A. Swartz, J. A. Hubbell, *Nature Mater.* **2007**, *25*, 1159–1164. *Exploiting lymphatic transport and complement activation in nanoparticle vaccines.*

[224]   S. N. Thomas, A. J. van der Vlies, C. P. O'Neil, S. T. Reddy, Sh. S. Yu, T. D. Giorgio, M. A. Swartz, J. A. Hubbell, *Biomaterials*, **2011**, *32*, 2194–2203. *Engineering complement activation on polypropylene sulfide vaccine nanoparticles.*

[225]   E. A. Mahmoud, J. Sankaranarayanan, J. M. Morachis, G. Kim, A. Almutairi, *Bioconjugate Chem.* **2011**, *22*, 1416–1421. *Inflammation responsive logic gate nanoparticles for the delivery of proteins.*

[226]   B. Yan, J.-Ch. Boyer, N. R. Branda, Y. Zhao. *J. Am. Chem. Soc.* **2011**, *133*, 19714–19717. *Near-infrared light-triggered dissociation of block copolymer micelles using upconverting nanoparticles.*

[227]   H. Zhao, E. S. Sterner, E. B. Coughlin, P. Theato, *Macromolecules*, **2012**, *45*, 1723–1736. *o-Nitrobenzyl alcohol derivatives: Opportunities in polymer and materials science.*

[228]   V. N. R. Pillai, *Synthesis.* **1980**, *1*, 1–26. *Photoremovable protecting groups in organic synthesis.*

[229]   D. Petrovi, R. Brückner, *Org. Lett.* **2011**, *13*, 6524–6527. *Deslongchamps annulations with benzoquinone monoketals.*

[230]   H. E. Gottlieb, V. Kotlyar, A. Nudelman, *J. Org. Chem.* **1997**, *62*, 7512–7515. *NMR chemical shifts of common laboratory solvents as trace impurities.*

[231]   H. R. Pfaendler, V. Weimar, *Synthesis* **1996**, *11*, 1345–1349. *Synthesis of racemic ethanolamine plasmalogen.*

[232] G. Mantovani, F. Lecolley, L. Tao, D. M. Haddleton, J. Clerx, J. J. L. M. Cornelissen, K. Velonia, *J. Am. Chem. Soc.* **2005**, *127*, 2966–2973. *Design and synthesis of n-maleimido-functionalized hydrophilic polymers via copper-mediated living radical polymerization: a suitable alternative to pegylation chemistry.*

[233] J. A. Heyes, D. Niculescu-Duvaz, R. G. Cooper, C. J. Springer, *J. Med. Chem.* **2002**, *45*, 99–114. *Synthesis of novel cationic lipids: Effect of structural modification on the efficiency of gene transfer.*

[234] C.-H. Liu, C.-Y. Pan, *Polym. Chem.* **2011**, *2*, 563–566. *A novel strategy for enhancing propagation rate of polystyrene grown from silica nanoparticles or carbon nanotubes.*

[235] Y. Kikuchi, J. Nakanishi, T. Shimizu, H. Nakayama, S. Inoue, K. Yamaguchi, H. Iwai, Y. Yoshida, Y. Horiike, T. Takarada, M. Maeda, *Langmuir* **2008**, *24*, 13084–13095. *Arraying heterotypic single cells on photoactivatable cell-culturing substrates.*

[236] C. Banzatti, A. Della Torre, P. Melloni, D. Pieraccioli, P. Salvadori, *J. Heterocycl. Chem.* **1983**, *20*, 139–144. *Derivatives of imidazo[5,1-c][1,4]benzoxazin-1-ones and related analogs.*

9 Appendices

9.1 List of Publications

Part of the results presented in this thesis has been adapted with minor modifications from the following publications:

1. E. Sokolovskaya, J. Yoon, A. C. Misra, S. Bräse, J. Lahann, *Macromol. Rapid Commun.* **2013**, accepted. *Controlled Microstructuring of Janus Particles Based on a Multifunctional Poly(ethylene glycol)*.

2. E. Sokolovskaya, S. Rahmani, A. C. Misra, S. Bräse, J. Lahann, **2013**, submitted. *Synthesis of a Multifunctional Poly(ethylene glycol) Derivative and its Use in Dual-Stimuli Responsive Microparticles*.

3. E. Sokolovskaya, S. Bräse, J. Lahann, in preparation. *Synthesis and On-Demand Gelation of Multifunctional Poly(ethylene glycol) Derivatives*.

The results contained in this thesis have been presented at the following conferences:

1. 3rd KIT PhD Symposium, Karlsruhe, Germany, 22.03.2012, oral presentation. <u>E. Sokolovskaya</u>, C. Friedmann, S. Bräse, J. Lahann. *Preparation of Functionalized Epoxides for Biofunctional Polymer Hydrogels.*

2. IUPAC MACRO 2012, Blacksburg, VA, USA, 24–29.06.2012, oral presentation. <u>E. Sokolovskaya</u>, S. Bräse, J. Lahann. *Multifunctional PEGs for Biomedical Application: Synthesis and Purification.*

3. 244th ACS National Meeting, Philadelphia, PA, USA, 19–23.08.2012, oral presentation. <u>E. Sokolovskaya</u>, S. Bräse, J. Lahann. *Synthesis and Purification of Homomultifunctional Polyethers and Their Block Copolymers.*

4. ERC Grantees Conference, Strasbourg, France, 22–24.11.2012, poster. <u>E. Sokolovskaya</u>, J. Yoon, S. Rahmani, S. Bräse, J. Lahann. *Multifunctional PEGs as a Synthetic Platform for the Preparation of Biomaterials.*

5. HYMA 2013, Sorrento, Italy, 3–7.03.2013, poster. <u>E. Sokolovskaya</u>, S. Rahmani, S. Bräse, J. Lahann. *Photoresponsive Poly(ethylene glycol)-based Materials for Biomedical Application.*

6. EPF 2013, Pisa, Italy, 16–21.06.2013, oral presentation. <u>E. Sokolovskaya</u>, S. Rahmani, J. Yoon, S. Bräse, J. Lahann. *New Multifunctional PEGs for the Design of Smart Drug Carriers and Hydrogels.*

9.2 Curriculum Vitae

Ekaterina Yur'evna Sokolovskaya

Personal Information

Place of Birth	15.07.1985
Date of birth	Dunaivtsi, Ukraine
Nationality	Russian
Marital status	Single, no children

Education

Sept.1991 – Jun.1999   Primary state school N30 of Vologda (Russia)

Sept. 1999 – Jun.2002   Secondary state school N1 of Vologda (Russia)

Sept.2002 – Jun. 2007   Master study of chemistry at the Lomonosov Moscow State University (MSU) in Moscow, Russia; degree: Master of Science in Chemistry. Diploma project in the research group of Prof. Dr. V. M. Demyanovich; topic: *"Synthesis of chiral 1,4-diols based on (S)-1-phenylethanol."*

Oct. 2007 – Oct.2009   Postgraduate student at the Lomonosov Moscow State University (MSU) in Moscow, Russia.

Oct.2009 – Dec.2009   Visiting Student at the Universidad Complutense de Madrid (UCM) in Madrid, Spain, in the research group of Prof. Dr. S. de La Moya Cerero; Topic: *"Development of camphor-based chiral ligands for asymmetric synthesis."*

Apr.2010 – Juni 2013   PhD work at the Karlsruhe Institute of Technology (KIT) in Karlsruhe, Germany, in the research group of Prof. Dr. J. Lahann/Prof. Dr. S. Bräse; Topic: *"Synthesis and Biomedical Applications of Multifunctional Poly(ethylene glycol)s."*

Jun.2012 – Aug.2012   Visiting Student at the University of Michigan (UM), in Ann Arbor, MI, USA, in the research group of Prof. Dr. J. Lahann; Topic: *"Synthesis of degradable nanoparticles by means of electrohydrodynamic co-jetting."*

9.3 Acknowledgements

It is a great pleasure for me to thank hereby my advisors Prof. Dr. Jörg Lahann and Prof. Dr. Stefan Bräse for the opportunity to work in their groups and for their friendly help and support during these three years of work on the thesis. I specially would like to thank Prof. Dr. Jörg Lahann for a very exciting and challenging topic and the possibility to carry out a part of the thesis work in his group in Michigan.

I would like to acknowledge funding from BioInterfaces Program at KIT as well as BioInterfaces International Graduate School established by the Helmholtz-association for providing interdisciplinary trainings and for bringing me together with many brilliant PhD students from all over the world.

I of course would like to thank all the people from the groups of Prof. Dr. Jörg Lahann and Prof. Dr. Stefan Bräse and other colleagues at KIT who created the warmest and friendliest atmosphere during these three years of work and readily helped me with all the issues I faced during this time. In particular I'm grateful to those of my colleagues with whom I worked most closely together and whose everyday help was extremely valuable: Dr. Florence Bally, Dr. Aftin Ross, Dr. Christian Friedmann, Dr. Leonie Barner, Dr. Dorota Jakubczyk, Dr. Caroline Hartmann, Dr. Hsien-Yeh Chen, Dr. Carlos Azucena, Domenic Kratzer, Artak Shahnas, Daniel Frank, Alessandro Donini. I thank Dr. Gerald Brenner-Weiß, Frank Kirschhöfer, Boris Kühl, Stefan Heißler and Dr. Maria Schneider for their great help with analytical methods.

Special thanks to Dr. Florence Bally and Dr. Aftin Ross for their most valuable corrections of the current thesis and suggestions on its improvement. A separate thank to Anne Meister for her advices concerning the thesis preparation.

I'm very grateful to my colleagues from the University of Michigan, with whom I worked together on some parts of the current thesis and who greatly helped me to integrate into the working environment there: Dr. Jaewon Yoon, Sahar Rahmani and Asish Misra.

I very much appreciate the constructive feedbacks and suggestions of the members of my Thesis Advisory Committee Prof. Dr. Jörg Lahann, Prof. Dr. Stefan Bräse, Prof. Dr. Christof Wöll and Dr. Pavel Levkin.

I'm also thankful to my scientific collaborators at KIT Dr. Ute Schepers und Dr. Alexander Welle.

I'm very grateful to Mrs. Astrid Biedermann and Mrs. Stefanie Sellheim-Ret for their kind help with the administrative processes and all other formal questions. In particular, I would like to thank Mrs. Astrid Biedermann for her help with my settlement in Karlsruhe.

My special and warmest thanks to my parents and my family for their permanent support and participation.

Of particular value for me is the care and help of my boyfriend Alessandro during the years of my work on the thesis. I'm greatly thankful to him for always being by my side.

Finally, I wish to thank my new friends in Karlsruhe who made my life here bright and happy and who also greatly supported me during the last three years: Priya Anand, Carmen Cardenal Pac, Christina Belenki, Simon Widmaier, Zung Choi and Elena Neverova. And of course I thank all my friends in Russia and other countries who always supported me from far away.